Perspectives on Engineering Uncertainty

William Nuttall · Edoardo Patelli · Ewan Smith ·
David Webbe-Wood · Simon Middleburgh

Perspectives on Engineering Uncertainty

Civil Nuclear Energy Safety and Efficiency

Foreword by Nawal K. Prinja

 Springer

William Nuttall
School of Electrical, Electronic
and Mechanical Engineering
University of Bristol
Bristol, UK

Edoardo Patelli
Department of Civil and Environmental
Engineering
University of Strathclyde
Glasgow, UK

Ewan Smith
Department of Mechanical and Aerospace
Engineering
University of Strathclyde
Glasgow, UK

David Webbe-Wood
London, UK

Simon Middleburgh
School of Computer Science
and Engineering
Bangor University
Bangor, Gwynedd, UK

ISBN 978-3-031-83253-6   ISBN 978-3-031-83254-3 (eBook)
https://doi.org/10.1007/978-3-031-83254-3

This work was supported by Bangor University.

This Springer imprint is published by the registered company Springer Nature Switzerland AG
The registered company address is: Gewerbestrasse 11, 6330 Cham, Switzerland

If disposing of this product, please recycle the paper.

Foreword: An Industrial Perspective on Engineering Uncertainty

Who doesn't want to know the unknown? In my 44 years of working in the nuclear industry, I've seen engineers and designers tackling the challenge of managing uncertainty. There are two types of uncertainty: one that can be reduced with more information and one that is natural variability and cannot be reduced by additional information. Methods have been developed to address both types, and with the advent of Industry 4.0, new digital technologies like artificial intelligence (AI) and machine learning (ML) are being applied. But, as with everything, there are two sides to the coin. While AI can be used for uncertainty quantification, there is inherent uncertainty in AI itself. The AI models and the data on which they are trained may contain uncertainty. Established methods exist to address data uncertainty and improve data quality and reduce bias, but reducing uncertainty embedded in an AI model can be challenging. One way to address this is to use an ensemble of models in decision-making rather than relying on a single AI model. Features can be added or engineered in a meaningful way. For example, the Reynolds number, a dimensionless quantity used in fluid mechanics to predict flow patterns, is calculated from density, velocity, characteristic length, and dynamic viscosity of the fluid. By calculating the Reynolds number for each set of data, analysts create a new feature that captures important information about the fluid regime. The key is to identify where uncertainty lies, understand its impact, and use the right methods to quantify it.

Addressing specific uncertainties has led to one of the most significant changes in the history of design codes and standards for engineering structures: the introduction of the EN Eurocodes [1, 2]. These new limit state codes replaced the traditional allowable stress codes, which specified a safe allowable stress or load limit that was usually a fraction of the material yield strength or the buckling strength, providing a factor of safety that accounted for uncertainty or variability. However, using a single factor of safety does not account for different types of uncertainties separately, which can result in either over-conservative or under-conservative designs. In the new codes, a structure is designed to reach a limiting state in which the maximum design loads are multiplied by a factor, and the strength (e.g., yielding or buckling) is reduced by dividing by a factor. These factors, known as Partial Safety Factors (PSF), depend on the uncertainty (scatter) in the demand/load and capacity/strength (caused by natural variability in material properties, geometry, and modelling). These factors are derived from statistical analysis and calibrated to ensure a certain level of reliability. Different factors for different types of loads and materials account for specific uncertainties, leading to more accurate and reliable designs. For readers keen to learn more please see [3]. Probabilistically calibrated Partial Safety Factors reflect real-world variability and uncertainties, leading to a more robust and reliable design methodology. With good understanding of the structural reliability methods, one can justify code compliance for different loads and materials which may not be possible with the traditional allowable stress codes.

The role of uncertainty quantification in designing the next generation of nuclear reactors cannot be underestimated. Decisions made by world leaders at COP 28 in Dubai emphasized that nuclear power must be part of the transition to net zero to mitigate the impacts of climate change. At the time of writing, in November 2024, it is expected that COP 29 in Azerbaijan will build on this global consensus and speed up the deployment of nuclear power that will provide further impetus to design new reactors and safety systems.

One of the biggest challenges in developing new products, particularly the next generation of nuclear reactors, is qualifying new materials and components. As the chair of the Senior Industry Advisory Panel of the Generation IV International Forum, a government-level cooperation agreement between 13 countries and 27 European Union members to develop the next generation of nuclear energy systems, it has come to my attention that the crying need of nuclear industry is to develop passively safe systems and new materials. Introducing new technologies and materials comes with the challenge of dealing with uncertainties. Engineers and designers aim to reduce the risk of failure by accounting for uncertainty in material behaviour under different conditions throughout the design's lifetime. Additionally, the nuclear industry is highly regulated, and introducing new materials requires navigating complex regulatory frameworks involving extensive documentation, testing, and approval processes. Codification of new materials requires several experimental tests, which can take decades and be costly. The existing approach is to train AI models with deterministic data that requires large number of physical tests to cover the natural variability in the material properties. An extra challenge for the nuclear industry is that there is not sufficient test data available for such a data centric AI approach. A recent feasibility study, PROMAP [4] has demonstrated that combining AI with probabilistic methods can help generate synthetic data to fill the gaps in the data. The solution is to combine artificial neural network (ANN) with Bayesian statistics and Interval Predictor Models to enhance the robustness of the data. This allows for the uncertainties from the sparse data and material variability to be accounted for and it allows to provide the necessary confidence associated with the predictions. AI-assisted uncertainty quantification can help fill gaps in sparse data that can optimise the number of material tests required. However, engineers need to be cautious about the level of confidence assigned to the synthetic data generated to fill gaps in the real data. Advanced manufacturing techniques like additive manufacturing (3D printing), powder metallurgy, and Hot Isostatic Pressing (HIP), combined with advanced inspection techniques like AI-assisted Phased Array Ultrasonic Testing (AI-PAUT), can improve the quality and reliability of nuclear components but may suffer from higher uncertainty during early stages of technology development. Digital technologies allow for real-time monitoring and adjustments, which can help reduce variability and ensure consistency. One such powerful technique is the Principal Component Analysis (PCA) that can reduce the dimensionality of datasets with large number of variables. It can reduce the dimensionality by identifying the most significant variables and patterns in data, making it easier to analyse and develop new materials and designs.

Data science and uncertainty quantification are intricately related. Advances in digital technologies that allow for the analysis of large volumes of data at speed are changing the way design and engineering are conducted. Historically, design engineers took an empirical approach, relying on their experience of what works and what doesn't. Nowadays, a more mechanistic approach is taken by studying the underlying science of a process or failure mechanism. But the future will be data centric. The nuclear industry has acquired volumes of real operational data which is simply archived. There is great opportunity to analyse the historic data to quantify uncertainty and optimize designs.

However, there are legitimate concerns expressed by nuclear regulators about the lack of explanation offered for decisions made using digital technologies like AI and ML algorithms that rely solely on data. Data can show what is changing but cannot explain why. AI is also prone to providing imaginary or impractical results, often called hallucinations. The way forward for engineers and designers is to add physics to AI models. A physics-informed neural network is a type of neural network that incorporates physical laws, often described by differential equations, which can be embedded directly into the network's learning process. In addition to data loss, physics loss is also added to ensure that the network adheres to known physical constraints and renders results that are practical and obey the laws of physics.

Designing complex plants is challenging, especially when faced with the dual challenge of reducing costs while increasing safety. In the UK, the nuclear sector deal issued by the government in 2018 called for a 30% reduction in new build costs by 2030. At the same time, after the Fukushima event, safety requirements have been toughened by the IAEA's Design Extension Conditions, which require plants to withstand multiple extreme hazards. Industry is responding to the challenge of reducing costs without compromising safety by changing their design approach. There is an initiative by a committee of international experts under the aegis of ASME to develop a new Plant Systems Design (PSD) code, which is a technology-neutral standard that provides a framework including requirements and guidance for design organizations. The PSD standard aims to bring three main changes: (a) integrate process hazard analysis early in the design stages, (b) incorporate and integrate existing systems engineering design processes, practices, and tools with traditional architectural engineering design processes, practices, and tools, and (c) integrate risk-informed probabilistic design methodologies with traditional deterministic design. The main feature and advantage of this new PSD code being developed is that it employs a systems-based approach to integrate design and safety by taking full advantage of employing uncertainty quantification during the design stages. Uncertainty quantification is fundamental in this approach, providing a systematic way to assess and manage the uncertainties inherent in complex plant designs.

It is a pleasure to see this publication by the eMEANSS project, which explains uncertainty and covers the important topic of tackling uncertainty. There remain several challenges in the nuclear industry to quantify uncertainty in reactor physics, structural integrity, and fuel performance. The knowledge and experience shared by expert authors in these areas in this book are valuable.

This book is not just a vital resource for nuclear engineers, but it also offers valuable insights that transcend the boundaries of the nuclear sector. By understanding and managing uncertainties, engineers and designers in aerospace, automotive, civil engineering, medical, and other fields can improve the reliability and safety of their designs and operations. The methodologies and tools discussed in this book will help build competence in tackling complex problems across various industries, ultimately leading to more resilient and efficient systems. Embrace the knowledge shared here to navigate the uncertainties in your own field.

As we look forward to the fifth industrial revolution (I5.0), that is going to focus on human-centric, resilient, and sustainable technologies, collaboration among engineers from various sectors in uncertainty quantification (UQ) will be pivotal. Engineers from different industry sectors can join forces to share knowledge and expertise to create robust UQ models tailored to a wide range of applications. Collaborative platforms for data sharing and integration can lead to better insights and more informed decision-making. A culture of continuous learning and collaboration can lead to a common set of competencies and skills. Most importantly, we must learn from the previous industrial revolutions, where standardisation and harmonisation often followed the industrialisation of technologies. For I5.0, let us aim to achieve 'standardisation before industrialisation,' paving the way for the development of smart cities where transport, energy, health, and education are seamlessly integrated into a cohesive, human-centric, resilient, and sustainable system. In this forward-thinking spirit, this book on uncertainty quantification offers invaluable insights and methodologies that will equip engineers to navigate the complexities of the future, ensuring that the advancements of I5.0 are both innovative and reliably safe.

Navigating uncertainty is both a science and an art. It is not just about statistical analysis using numbers but also about how to use the results in decision-making.

This book will help all those decision-makers who must deal with the known and the unknown.

Prof. Nawal K. Prinja
Prinja and Partners
Stockport, UK

References

1. Brettle ME (2009) Steel building design: introduction to the Eurocodes. The Steel Construction Inst., Publication P361. https://www.steelconstruction.info/images/0/0a/SCI_P361.pdf
2. Eurocodes: building the future. https://eurocodes.jrc.ec.europa.eu/
3. Rationalisation of safety and serviceability factors in structural codes, CIRIA Report 63, 1977
4. Probabilistic artificial intelligence prediction of material properties (PROMAP) for nuclear reactor designs. In: Lye A, Prinja N, Patelli E (eds) Proceedings of the 32nd European safety and reliability conference (ESREL 2022). https://www.rpsonline.com.sg/proceedings/esrel2022/pdf/S24-02-306.pdf

Acknowledgements The research, writing, and publication of this book have been supported by the Engineering and Physical Sciences Research Council UK via a grant entitled: *Enhanced Methodologies for Advanced Nuclear System Safety (eMEANSS)*. The grant has reference: EP/T016329/1.The authors thank the EPSRC for this support.

The authors are most grateful to the entire eMEANSS team for their input and support to this project at various stages. eMEANSS has been a joint project involving researchers from the UK and India. This book is an output from the British side of the collaboration. The authors extend their special thanks to James Marrow, Geoff Parks, Chris Truman, and Karl Whittle, for their reading of the draft manuscript and for their most helpful comments. The authors also thank Mike Bluck, Robin Grimes, and Simon Walker for their help with other aspects of the EMEANSS project on the UK side. We stress that these various kind helpers bear no responsibility for the book presented here.

Special thanks are due to Dr. Nawal K. Prinja. Dr. Prinja made several important and profound contributions to the understanding of the eMEANSS team via his participation in key events and meetings during the research project. Dr. Prinja has kindly provided a foreword for this book and for that the authors are most grateful.

The authors are also most grateful for the input provided by eMEANSS collaboration with Bhabha Atomic Research Centre in India including particularly Anurag Gupta and Imran Khan. They were generous hosts and kind contributors to the eMEANSS research project. We are also most grateful to Umasankari Kannan who was so important in the original development of the ideas that became the eMEANSS project.

It is important to stress that all those who offered friendship and advice are not responsible in any way for the content of the book presented here.

The authors are most grateful to our publisher Springer for making it possible to publish our work open access. We also extend our thanks to our commissioning editor Anthony Doyle and to all those involved in book production including Amudha Vijayarangan and Anu Peter.

Competing Interests The authors have no competing interests to declare that are relevant to the content of this manuscript.

Disclaimers

Liability: The authors and the publisher cannot assume responsibility for the validity of all materials or for the consequences of their use. This book contains no advice or guidance and should not be used as the basis of any investment or other decision.

Trademarks: Product or corporate names may be trademarks or registered trademarks, and are used only for identification and explanation without intent to infringe.

Images: Efforts have been made to establish and contact copyright holders for all images presented in this book. We are most grateful to all the various rights holders who have kindly granted permission for reproduction. Despite our best endeavours, there may be instances where the rights of third parties have been overlooked. In such cases, we apologise and we ask that rights holders make contact and we will endeavour to resolve matters.

Contents

Chapter 1
Introduction

This book arises as a consequence of a UK government funded research grant forming part of the UK-India Civil Nuclear Research Partnership. This book is produced by some of the UK researchers in that endeavour. The authors express their thanks to all involved in the project (see Acknowledgements). The research project is formally known as: enhanced Methodologies for Advanced Nuclear System Safety, but it is known by the team as "eMEANSS" and more about the wider project can be found at the project website: https://nubu.nu/emeanss/.

The research project involved four work packages:

- Reactor Physics
- Materials Science of Nuclear Graphite
- Nuclear Fuel Performance
- Overarching Synthesis and Dissemination.

This book forms an output of the fourth work package. It is important to stress at the outset that this short book is not a textbook. It presents an overview of linked concepts and issues. It is intended that it will allow an interested reader to gain an appreciation of current issues in the management of engineering uncertainty and the book provides pointers for the reader to allow for further learning. The book does not set out to teach the methods and techniques to which it will refer. As such, we hope it can be a primer for those new to the field.

The eMEANSS project brings together technical experts familiar with most of the key issues underpinning the safety of a complex nuclear power station. Such considerations include in-reactor fuel performance, the cladding of the nuclear fuel, the moderation of the nuclear chain reaction and the structural integrity of the reactor as a whole. The eMEANSS team was also supported by experts involved in the conversion of nuclear fission energy to electrical energy via steam generation and use. Such considerations involve thermal hydraulics. The expertise brought together for the eMEANSS project is deployed to address specific problems of current research interest to the global nuclear engineering community. As such, eMEANSS research

© The Author(s) 2025
W. Nuttall et al., *Perspectives on Engineering Uncertainty*,
https://doi.org/10.1007/978-3-031-83254-3_1

is devoted to matters affecting both normal operations of a reactor and severe nuclear accident scenarios. In this book we shall adopt a medical metaphor and refer to this distinction as being around 'chronic' and 'acute' considerations.

The work undertaken across the eMEANSS research project considers a range of fission reactor designs including pressurised water reactors and graphite-moderated high temperature reactors—as may be deployed in the future for industrial process heat applications. It is not the purpose of this book to share the latest research insights relating to the key challenges in nuclear engineering uncertainty management for these specific technologies. Rather, we seek to illustrate some more general philosophical and practical lessons emerging from the nuclear engineering community as it looks to develop a fourth generation of nuclear energy systems. The immediate future of nuclear fission energy involves a range of exciting and innovative reactor concepts including molten salt reactors, fast reactors, and high temperature gas cooled reactors. In addition, there is renewed interest in fusion energy (Nuttall et al. in press). Meanwhile innovation continues to push forward the pressurised water reactor (PWR) concept, with particular innovation occurring in designs for improved buildability and operational flexibility—that is via the move to small modular reactors (SMRs).

Specifically, the eMEANSS project is investigating the use of uncertainty modelling in fuel performance, structural integrity and reactor physics studies. It aims to make use of developments in uncertainty modelling to improve our knowledge and understanding of these uncertainties and how they interact.

As a large engineered system, such as a nuclear reactor, is manufactured and operated, the inherent complexities build up, and key uncertainties are propagated from one level of the technology to another. Without a knowledge of how the uncertainties build up, it has been necessary to apply large amounts of conservatism in both design and operating practices. However, improving holistic knowledge of system uncertainties and how they propagate would allow nuclear reactors to be operated more efficiently and indeed more safely. The eMEANSS project has been focussed on that task.

It is reasonable to ask: in what way is nuclear reactor safety distinct from the safety issues relating other safety critical advanced technologies, such as in aviation? To be sure both aviation safety and nuclear reactor safety involve highly multidimensional data sets—see Sect. 3.3. Aviation and nuclear power differ, however, in terms of the level of accumulated experience and indeed in the amount of operating equipment at any instant in time. While roughly 100,000 flights take place every day, the number of operational nuclear power reactors is only around 400. Much of the global story of aviation safety has been a robust process of learning from accidents and near miss incidents. The amount of such data is not insignificant despite the high levels of safety in the aviation industry simply because so much activity is taking place. For nuclear power, in contrast, the opportunity to learn from bad experiences is very limited because the number of incidents is so very low, in part because the total number of power reactors is itself low. This allows us to introduce a conceptual point of difference between nuclear power safety and most other safety critical industries, including aviation. Nuclear power safety analysts must live with the reality that the real world data sets are sparse. Of course, one can simulate, and digital twins (see Sect. 6.2) are very helpful in that regard, but historically, at least, real world experience matters.

The data may be sparse, but new technologies present new opportunities. For example, innovations in computer modelling, artificial intelligence and supercomputers provide potentially improved approaches to developing new materials and understanding key uncertainties. For nuclear power these innovations can be adopted and brought in from sectors where they have been tested in a very large number of instances. Additionally, developments in computing and information technology typically allow significantly more calculations to be carried out and they enable the improved extrapolation of data beyond measured values. Such improved knowledge can improve safety. It also has the potential to improve the economic competitiveness of nuclear power.

As for an illustrative example: the current knowledge of the behaviour of concrete and steel structures is good. However, close to the reactor core, where temperatures and radiation levels are high, our knowledge becomes less certain. In the past, to be able to meet the requirements of an anticipated 60–70-year operating lifetime, uncertainty has been compensated for by over-engineering. Improved knowledge of uncertainty will allow improvements in the manufacture and operation of reactors while maintaining safety standards.

It is not our intention in this book to survey the nuclear landscape, such an overview can be found elsewhere (Nuttall 2022). In this book we aim to consider ideas both philosophical and practical, and in order to do that we focus on a single reactor concept as a device through which we can simplify the exposition of ideas in this book. After much deliberation, we have focussed this book on the British Advanced Gas-cooled Reactor (AGR) concept deployed in the UK, and only the UK, in the second half of the twentieth century. We choose to focus this technology type for two main reasons:

- As with other gas cooled reactor concepts, such as the older UK Magnox design, the AGR concept has a clear distinction between the fuel, moderator and coolant. As such it is logically simpler to appreciate the role played by different elements of the design. For an introduction to the concepts of fuel, moderator and coolant the reader is referred to the glossary at the end of the book and for more information to a nuclear engineering textbook such as that authored by Professor Malcolm Joyce (Joyce 2017).
- The AGR is not a commercial proposition for new build in the twenty-first century. It is not our intention with this book to comment on the merits, or otherwise, of any nuclear technologies available on the market today. Our focus is merely to explain how uncertainties can be viewed and managed in modern complex engineering projects, such as nuclear power stations.

The new and emerging methods discussed in this book are intended to increase the confidence in the assessment of safety of nuclear plants (and potentially in other safety critical industries). By improving the quantification of, and combination of, uncertainties in the underlying processes and parameters. This will lead to increased confidence in the safety assessments.

In Chap. 2 we look back at the history of the handling of uncertainty in engineering practice.

In Chap. 3 we introduce key concepts relevant to engineering uncertainty and provide examples whereby modern innovations, especially in computing, have made possible new analytical techniques.

In Chap. 4 we introduce the UK Advanced Gas-cooled Reactor (AGR) as a canonical nuclear reactor engineering concept. We use the AGR as a vehicle by which various nuclear engineering points can be better explored.

In Chap. 5 we illustrate some key insights into uncertainty handling in engineering with reference to the reactor design concept introduced in Chap. 4.

In Chap. 6 we point to emerging ideas and approaches as explored by the authors and collaborators in a nuclear engineering context, noting that these techniques may be of wider utility.

In Chap. 7 we summarise and offer some closing conclusions.

Acknowledgement This work was supported by the Engineering and Physical Sciences Research Council, UK via a grant entitled: *Enhanced Methodologies for Advanced Nuclear System Safety (eMEANSS)*. The grant had reference: EP/T016329/1. The authors thank the EPSRC for this support.

References

Joyce M (2017) Nuclear engineering: a conceptual introduction to nuclear power. Elsevier Science & Technology. ISBN: 9780081009628

Nuttall WJ (2022) Nuclear renaissance: technologies and policies for the future of nuclear power, 2nd edn. Routledge. ISBN: 9780367478070

Nuttall WJ, Konishi S, Takeda S, Webbe-Wood D (in press) Commercialising fusion energy: how small businesses are transforming big science, 2nd edn. Whittles Publishing, Dunbeath Scotland

Chapter 2
Thinking About Uncertainty

Uncertainty links to the science of measurement and it plays a key role supporting the notion of prediction in science and engineering. Measurement is a scientific activity, but it is also a key concept in metaphysics and research philosophy. Prediction of future events relies upon theories concerning the laws of nature and models of the world in which we live.

To illustrate these arguments, let us consider a scientific question of the early modern era—the prediction of tides. Tides in the open seas and oceans of the world are the consequence of lunar and solar gravitational fields influencing the liquid surface of planet Earth. Imagine for a moment that we want to predict the times of high tides at some specified place. For the sake of argument let us select Lisbon, Portugal. In principle there are two ways to approach this problem. The first is purely empirical and grounded in the science of measurement and the other is more theoretical in nature.

The empirical approach would involve the recording of tide times in Lisbon over a long period—ideally more than one year. From this data a sufficient time series can be obtained to allow the data to be fitted with mathematical functions selected solely for their ability to match to the observed data. There is no need for such an approach to have an underlying theoretical reasoning in support of the fitting curves selected, in order for the approach to be effective. Once such fits are available, the resulting understanding can be used inductively to infer the times of future high tides in this particular place.

In extremis the alternative theoretical approach requires no experimental observations of tides whatsoever, although the approach would rely on a set of wholly independent astronomical observations, presumably obtained previously by generations of astronomers. In the theoretical approach, one starts with a physical model of the Earth-Moon system, noting that the gravitation effects of the Moon at the Earth are larger than those of the Sun. The theory of tides is rather counter intuitive in that high water is found at those points on the surface of the Earth that are closest and furthest from the Moon. These realities and the fact that the Earth rotates on its

© The Author(s) 2025
W. Nuttall et al., *Perspectives on Engineering Uncertainty*,
https://doi.org/10.1007/978-3-031-83254-3_2

axis and that separately the Moon rotates around the Earth ensure that the earth's tide interval is not 12 h, but rather 12 h and 25 min (NOAA 2024). With such a simple two body astronomical model it is in principle possible to construct an approximate tide table for any coastal point on planet Earth. A more sophisticated theoretical approach would bring in the gravitational effects of the Sun, and hence yield a tricky three-body problem, but with the help of modern computers an improved tide prediction algorithm could be assembled. In this way tides for Lisbon might be predicted.

One might intuitively take the view that the theoretically grounded approach is superior in its predictive powers for Lisbon. The approach is certainly more intellectually more demanding, but it is probably not more accurate in its forward predictions. The benefit of the theoretical approach is its generalisability. The empirical approach really only works for the place where the measurements were made.

Philosophically this little story relates to a distinction between inductive and deductive reasoning. The approach that we have so far called empirical is inductive and, as we said, it can work well for a specific site. To predict tide behaviours of any coastal location, and with minimal information from that place, one needs to adopt a deductive approach. We previously called this 'theoretical' because it was grounded in the theory of planetary astronomy.

Of course, real science is a combination of both the inductive and the deductive. Models are built from deductive hypotheses and tested against empirical data or observation. In this way models are tested, validated and improved.

2.1 Safety Margins Matter

We have described the concept of a safety margin in the glossary of this book with the words: *"an intentional overestimation of certain parameters to ensure that a design or system can tolerate unexpected variations or uncertainties without compromising safety or functionality"*. In a book dedicated to issues of engineering uncertainty it is hardly surprising that we see the definition of a safety margin in terms an ability to tolerate variations or uncertainties. The UK Office for Nuclear Regulation also emphasises the importance of uncertainty when it states (when considering claims for the reliability of a nuclear installation) in ERL 4 of its Safety Assessment Principles: *"the safety case should include a margin of conservatism to allow for uncertainties"* (Office for Nuclear Regulation 2020). That phrase talks to a set of ideas and approaches that taken together comprise a set of safety margins. These components of the nuclear power station safety margin include: beyond design basis margin, lifetime margin, margins on toughness, shut down margin, and sub-criticality margin. The full set of safety margins for a nuclear power station surely has even more components than those described specifically here. Taken together they form the safety margins.

Sometimes one encounters a definition for the margin of safety (MS) based on the ratio of the strength of a structure to its design requirement. As Waqqas Ahmad helpfully observes: *"One usage of margin of safety is as a measure of capacity*

[...] (Ahmad 2024). The other usage of MS is as a measure of satisfying design requirements (requirement verification)." He goes on to explain: *"Margin of safety can be conceptualized [...] to represent how much of the structure's total capacity is held "in reserve" during loading."* His position is summarised with the words: *"there are two separate definitions for the margin of safety, so care is needed to determine which is being used for a given application".*

We align with Ahmad when we see things in terms of capacity. Mathematically we see that as a definition in terms of the number of standard deviations that separate the (standardised) design capacity of a component/structure from the (standardised) demand or load. As Ahmad alludes, this is well known concept in structural design. As for the link between safety margins and uncertainty the IAEA has given this much thought as the considerations apply to nuclear power, see for example IAEA TecDoc 1332 (IAEA 2003). As the introduction to the IAEA TecDoc explains: *"For practical purposes, the safety margin is usually understood as the difference in physical units between the regulatory acceptance criteria and the results provided by the calculation of the relevant plant parameter".*

One example would be the maximum tolerable accident temperature for nuclear fuel. An engineering assessment of the maximum temperature that must be tolerated might yield a result consistent with the "safety limits" line shown in Fig. 2.1. Every measurement of temperature involved in such considerations including the temperature of the reactor entering an accident scenario will come with an uncertainty, be that due to sensor noise/degradation, natural variations or, more likely, a combination of both. Such considerations take us to the dashed horizontal line in Fig. 2.1. One conception of safety margin is the difference between that dashed level and the regulatory acceptance criterion (itself set slightly below the actual safety limit). Another concept of safety margin leads us to the dark blue line in Fig. 2.1—the positioning of the dark blue line incorporates a level of conservatism beyond the considerations of uncertainty. In Fig. 2.1 one can see two different logical approaches to the concept of safety margin, which while broadly similar, do differ in detail.

Related to the concept of the Safety Margin is the idea of a Safety Factor. Simply put Safety Margin is expressed in absolute terms while a Safety Factor is expressed as a proportion. For example, a steel plate that is calculated as requiring a thickness of 10 mm might be judged to require a Safety Margin of 1 mm—in that case the Safety Factor would be 10% (1 mm/10 mm). Assuming a linear relationship between geometry, physical loading and the safety limit at hand, this would allow the part to sustain a loading level 10% above the pre-calculated maximum design limit.

In practice, single value safety factors are however rarely used these days in complex, safety critical contexts. As Dr Nawal Prinja once reminded the eMEANSS researchers: "The Factor of Safety is dead". The fixed, somewhat arbitrary, deterministic approach illustrated in Fig. 2.1 turns out, in practice to be less useful than a more probabilistic conception of the safety margin grounded even more fundamentally in notions of uncertainty. Rather than fixed values of key parameters it is more helpful to work with probability distributions which depict the variation we can expect to see in the real-world. Figure 2.2 illustrates the two realities which would be said to have

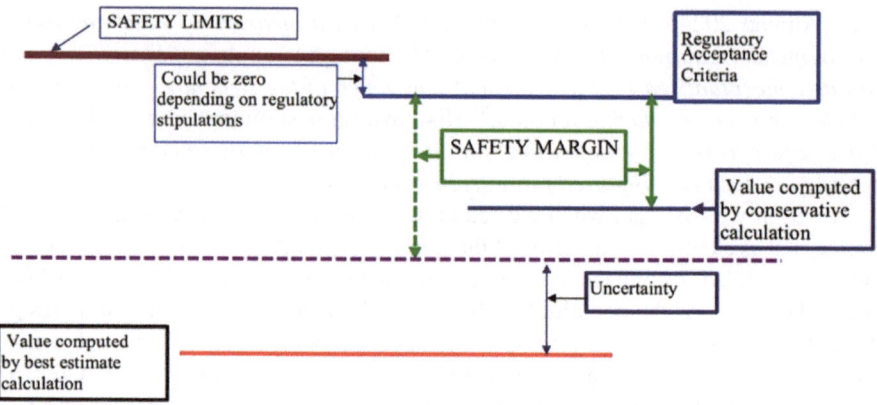

the same factor of safety (or reliability)—that is, the mean values of the two distributions are separated by the same amount in each case. Both scenarios are normalised. In the lower case there is little or no overlap of the two distributions, while in the upper case the overlap is significant. The large amount of overlap implies that the risk of failure is significant, while the small amount of overlap suggests a smaller risk of failure—and yet, they share the same factor of safety!

Figure 2.3 adjusts Fig. 2.1 to accommodate the use of probability distributions in place of single point values. Now, the uncertainty range is defined by the variance in the probability distribution of the quantity of interest (the range may be defined by 95th or 99th percentiles of the distribution or similar). The Safety Margin now becomes the probabilistic safety margin, the distance from the uncertainty range to the regulatory limit of interest.

Fig. 2.2 Same safety factor but different reliability. From: https://ntrs.nasa.gov/api/citations/20120001369/downloads/20120001369.pdf. US Government (NASA)

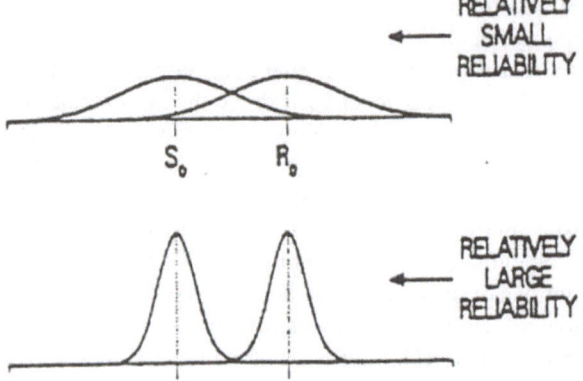

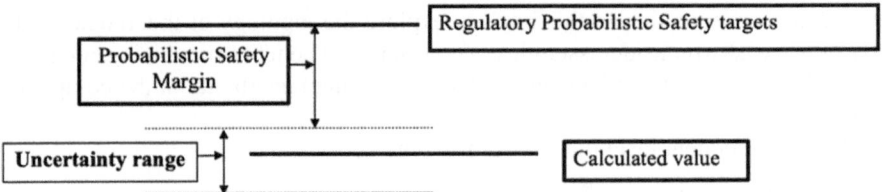

Fig. 2.3 The logical basis of a probabilistic safety margin. Published with kind permission of the IAEA, Source and copyright holder: International Atomic Energy Agency, Safety Margins of Operating Reactors Analysis of Uncertainties and Implications for Decision Making, IAEA-TECDOC-1332, IAEA, Vienna (2003). All rights reserved

On paper, this sounds good. We can determine safety margins calculated with the real uncertainty we can expect in operation, rather than using potentially arbitrarily defined safety factors. But what can uncertainty look like? What shapes can it take? What if we don't have enough data to create a nice probability distribution? Are there different types of uncertainty? Let us explore all this in the next subsection.

2.2 Uncertainty and Engineering Safety

Thus far in this book, and indeed throughout this book, we focus on aspects of uncertainty that are defined within a given snapshot in time—uncertainty in temperature, thickness, hardness or energy conversion efficiency etc. As such our focus has not been on the time evolution of uncertainty as one thing affects another moving forward in time. It is, of course, conventional in nuclear safety assessment to consider consequential processes occurring over time linking to concepts such as causality and consequence. Such logics link to established considerations in nuclear engineering, such as:

- Decision trees and fault trees
- Probabilistic Risk Assessment
- Cascading failure.

Without wishing to diminish the importance, or value, of such temporal considerations, in this book we will give emphasis to uncertainties that exist at a fixed point in time. That is, those that have a meaning within a snapshot in time can be a rich functional[1] of various uncertainty functions governing the underlying processes and parameters that are incorporated into safety assessment, even before notions of time dependency, causation and consequence are introduced. For readers seeking insight into the importance of causality and cascading risks we recommend the work of Alexander and Pescaroli (Alexander and Pescaroli 2019).

---

[1] Simply put, a functional (mathematical noun) is a function of functions. More precisely it is a real-valued function on a vector space. Functional analysis forms part of the calculus of variations.

It is not our intention in this book to explore the topology of theoretical vector spaces, we wish to consider issues in concrete terms of benefit to practicing engineers. Let us imagine a team and a team leader confronting uncertainty at the component level.

At the component level the issues might include:

- First, what are the boundaries of the model? The model has a range of validity, that is itself an uncertainty
- Second, the data, as used to build the model, or more precisely what the team chooses to feed the model, also comes with uncertainty
- Some parameters are more sensitive than others, with sensitivity a key dimension of uncertainty at the component level.

Typically, the team would find the deterministic values, or the most probable values, and run the team's model based on those values. This leads a single point value output. That is one single output parameter which, if the recipient is lucky, comes with a simple post-hoc error assessment. A more sophisticated analysis, however, might repeat the model run multiple times. Such an approach might be based on a stochastic variation of key parameters (such as in the Monte Carlo approach). In this way one can start to appreciate sensitivities with respect to key parameters, but even then—on the output side one is always dealing with a single value report as for inclusion at the whole system level.

In this book we suggest that those responsible for project management properly define, characterise, and propagate the uncertainty throughout the system and all its modelled elements. The component level project team must pass that kind of information to an analyst at a higher (second) level, who will couple that result with their own system analysis. As alluded to earlier, in mathematical terms the whole system management should be dealing with a functional (i.e. a function with inputs that are themselves functions, rather than single point values).

As we urge project engineers to keep track of uncertainty propagation, we are conscious that the current usual reality is very far from that situation. In Fig. 2.4 we illustrate that, based on the current level of information and knowledge, a set of probability distributions can be used to represent a parameter of a component (the information coming from different assumptions or modelling approaches). The set of all the distributions should be propagated without any mixing (averaging of them) (Gray et al. 2021). A real nuclear power plant will usually combine many thousands of such parameters, with their associated uncertainty characterisations, into an overall performance measure, and that is before we acknowledge the complication of multi-dimensionality of the data sets.

Despite the reality that nuclear experience-based data sets are typically sparse (see discussion in Chap. 1) it may be possible to generate an overall sense of a performance distribution for a modular element of a nuclear power station or even, perhaps, for the power station as a whole. If one were to be able to generate a curve representing whole system behaviour then, as things stand, typically one has no way of unpacking that overall understanding so as to be able to appreciate its component parts, as illustrated stylistically in Fig. 2.4. It means that typically one has no way

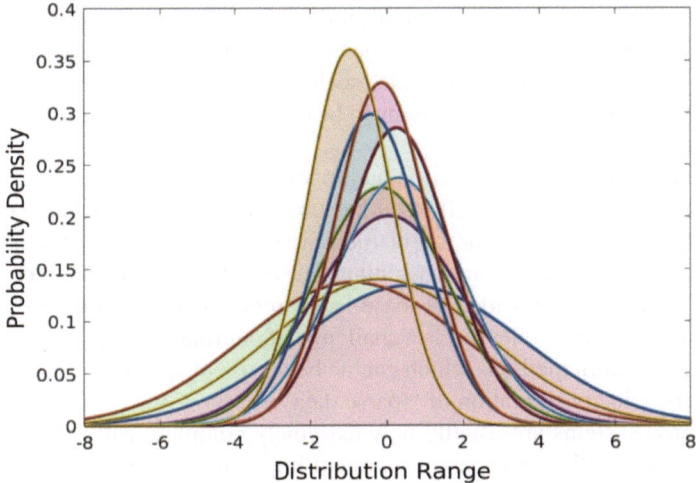

Fig. 2.4 The uncertainty in the key safety properties of a highly engineered system, such as a nuclear power station, is a consequence of innumerable underpinning distributions, as illustrated in stylised form here. *Source* Edoardo Patelli—with kind permission—all rights reserved

properly to identify the real response and therefore the entire envelope of the response should be used wherever possible.

In the last paragraph we considered modular elements of a nuclear power plant. A similar logic can apply to the various specialist teams the contribute to the engineering of a nuclear power station. In engineering organisations with groups of teams, the teams are under pressure to minimize uncertainty. There may be a parameter that one team spends a lot of money and effort to minimise, when in fact it's not sensitive for the operation of the whole engineering system. At this point we see that the issues of concern relate to communication breakdown and whole system sensitivity to input parameters and exogenous inputs.

Traditional engineering uses Safety Margins (see Sect. 2.1) in design to compensate for the inherent variability and uncertainty in modelling and simulation. The conventional approach to assessing the effects of the input variables is to assume or estimate "worst case" values and determine the design output variable accordingly, by standard engineering methods.

Engineering design provides a systematic approach to delivering a product, whether physical or data-based. It not only concerns the performance of the product in its intended use, but it also addresses the key challenges regarding its synthesis and the provenance of the raw materials and data, and the uncertainties that lie within them. Prototyping in engineering has developed too, moving from 1:1 models to scaled representative systems. Originally this would involve small prototype devices constructed, as far as possible, from the same materials to be used in the final commercial product, but now scaled prototyping uses computational methodologies. With this approach comes an increase in scientific burden of providing the necessary correlations, logical dependencies and models. It is essential that these IT based representations match to physical realities and that they can be validated, as such. Increasingly, however, in recent decades such an approach has become possible and trustworthy, and this allows for systems to be optimised in a more cost-effective and timely manner.

In addition to design for operation and manufacture, the nuclear industry has an obligation, and is proud of the efforts it has taken, to understand what is to be done with its waste, whether that be spent fuel, arisings from reprocessing streams or structural materials from decommissioning. The development of engineering design principals sufficient to consider all these facets (construction, operation, waste and decommissioning) as a collective is a challenge of our times. That challenge is being addressed by modern methods which enable better use of sparse and stochastic data. We shall return to consideration of 'sparse data' in Sect. 6.4.4.

Engineered systems are usually now extremely complex, with many interdisciplinary and intersecting portions of a system impacting on one another. The traditional chief engineer role, who can understand and sign off on the safety report is no longer a tractable and singular person. Instead methods used to define engineering design need to be tested and understood, and a holistic approach to uncertainty and risk needs to be baked into the methods. The chief engineer of the future will have a good overview. Their viewpoint will be enabled by technology and be founded upon different logical principles. In this way, they should be au-fait with prediction, modelling and most importantly the uncertainties that surround them on a day-to-day basis.

Acknowledgement This work was supported by the Engineering and Physical Sciences Research Council, UK via a grant entitled: *Enhanced Methodologies for Advanced Nuclear System Safety (eMEANSS)*. The grant had reference: EP/T016329/1. The authors thank the EPSRC for this support.

References

Ahmad W (2024) Factor of safety and margin of safety. https://www.mechanical360.net/updates/factor-of-safety-and-margin-of-safety/. Accessed Date 16 Oct 2024

Alexander D, Pescaroli G (2019) What are cascading disasters? UCL open environment. https://doi.org/10.14324/111.444/ucloe.000003

Gray A, Davis A, Patelli E (2021) Uncertainty propagation in SINBAD fusion benchmarks with total Monte Carlo and imprecise probabilities. Fusion Sci Technol 77(7–8):802–812. https://doi.org/10.1080/15361055.2021.1895667

IAEA (2003) Safety margins of operating reactors: analysis of uncertainties and implications for decision making. Vienna. https://www-pub.iaea.org/MTCD/Publications/PDF/te_1332_web.pdf

NOAA (2024) Tides and water levels. https://oceanservice.noaa.gov/education/tutorial_tides/tides05_lunarday.html. Accessed Date 10 Aug 2024

Office for Nuclear Regulation (2020) Safety assessment principles for nuclear facilities, 2014 edn, Revision 1. Bootle 2020. https://www.onr.org.uk/publications/regulatory-guidance/regulatory-assessment-and-permissioning/safety-assessment-principles-saps/

Chapter 3
Unpacking Uncertainty

In this chapter we explore uncertainty as a philosophical concept, before converging on those aspects of greatest relevance to engineering and, in particular, large-scale safety critical engineered systems such as nuclear power stations. We will argue that quantum indeterminate measurement, and deterministic chaos lie largely outside the scope of our interest, but nevertheless both concerns lie at the heart of a key consideration—the role of classical determinism. We shall conclude that it is not valid to conclude that a deterministic system can necessarily be modelled to the point that uncertainty can be eliminated. By that, we mean that if initial conditions are well understood and all interactions within the system are properly captured, one can still not safely assume that the future state of the system can be predicted with certainty.

When we suggest that complete certainty is not an appropriate goal for engineering design, we equally do not wish to imply that a deeper understanding of uncertainty in its various dimensions and levels cannot lead to a reduction in uncertainty. For example, it can be the case that if the boundaries of uncertainty are better known, then in practical terms uncertainty itself may be regarded as being more limited. In this chapter we intend to walk through these arguments in a stepwise manner.

3.1 Uncertainty and Determinism

When thinking of engineering uncertainty, and if guided by concerns of practical relevance, then uncertainty can be defined as one's inability to measure the outcome of events which are of interest. At this stage we shall deal with the manifest observation that uncertainty of this type can arise due to a limitation in our knowledge about the system of interest. For example, we may be lacking information about the underlying processes on which the technology relies, we may have knowledge gaps (known and unknown—more about that later), there may be system complexity (in

© The Author(s) 2025
W. Nuttall et al., *Perspectives on Engineering Uncertainty*,
https://doi.org/10.1007/978-3-031-83254-3_3

the formal sense understood in complexity science—including for example emergence), modelling errors (including omissions) and a lack of sufficient computational capability. It is the last concern that has greatly reduced in recent decades, despite the fact that problems being put forward for computational assessment have to some extent expanded to fill the capability available. Typically, these knowledge-based sources of uncertainty do not each contribute in isolation. Uncertainties of this type are usually found in combination with one another—for example weak computational capability in the past may have led to incomplete modelling with excessive use of approximation and assumption.

3.1.1 Uncertainty and the Quantum

In quantum mechanics, Heisenberg's principle of indeterminacy (uncertainty) states that some physical quantities such as the position and momentum of a particle, cannot simultaneously be known with arbitrary precision (Heisenberg 1927). In other words, knowing the exact position of a particle, such as a photon or electron, induces a limit upon the maximum accuracy of the momentum that can be estimated and vice-versa. We mention this quantum indeterminacy simply to make the point that the universe is not intrinsically deterministic.

3.1.2 Uncertainty and Chaos

At the macroscopic level, although the effect of the aforementioned Heisenberg uncertainty principle can be negligible at the engineering level, a full knowledge of the system of interest is still impossible to obtain. It is the wider idea that full knowledge is generally unobtainable for real systems that we shall consider further. Despite the underlying processes of most physical systems being well understood, their mathematical modelling always includes some degree of approximation or simplification to make the model tractable. One source of apparent randomness in classical deterministic systems can be 'chaos'.

The science of deterministic chaos is profound and even somewhat counter-intuitive. This does not invoke quantum physics, but it reveals emergent behaviours that are to all intents and purposes—unknowable.

A canonical example is revealed by the simply expressed logistic equation from population ecology:

$$x_{t+1} = k\, x_t (1 - x_t)$$

Here k is the 'control parameter' and in ecology it relates to the net birth/death rate.

This assumes a normalized population, such that '1' is the maximum possible population. This forces $0 < k < 4$.

The interesting thing is to choose values for k and then to iterate this equation.

If $k < 1$ then as $t \to$ infinity, $x_t \to 0$ for all x_0. The point '0' is an 'attractor' in that case.

For k in the range 1 to 3 the attractor increases from zero to around 2/3. For k in this range the attractor can be predicted from the value of k. For $k > 2$ the convergence is oscillatory.

For $k > 3$, but close to 3, there is not one attractor x^*, but two. A bifurcation is said to have occurred. Each iteration the value of x jumps between two values. At large iterations the two attractor values become clear. The attractor values are also dependent on k.

For k slightly larger than 3, the first few bifurcations in x^* are occurring. This is not the chaotic regime. At first there are two attractors, then 4, then 8—the number doubles quickly.

For $k > 3.57$ the system is in the chaotic region, the number of attractors x^* is no longer doubling as k is increasing—the number of attractors is at least 16 and is rising very fast with k.

The chaotic regime is $k > 3.57$ in this region the number of attractors is vast and the value of these attractors x^* varies wildly with k. It is impossible to infer k from x^*.

Only for $k = 4$ can x^* take all values in the range 0–1.

For $k < 4$, there are always limits to the extent of the chaos.

A chaotic system is one that appears to be disordered and randomly interacting. It is sustained and evolves over extended periods of time. Key to a chaotic system is its governance by distinct mathematical rules that are deterministic and non-linear. Importantly chaos does not rely on quantum or other non-classical physics. The determinism of late nineteenth century physics is sufficient for chaotic behaviours to occur (Gleick 1997; Williams 1997).

It is impossible to obtain a full knowledge of a chaotic system. Despite the underlying processes of most physical systems being well understood (that is the underlying mathematics may be well understood), any mathematical modelling always includes some degree of approximation or simplification to make the observed behaviour of the model tractable.

The best known, and indeed the canonical, example of simple underlying mathematics giving rise to chaotic system behaviour is the logistic regression, see Fig. 3.1.

Interesting, as a practical generalisation, in the chaotic regime it is not possible to infer from the observed outcome (the attractor) what the underlying causative mathematics is. Such considerations take us to concerns that lie beyond the scope of this book and which would also bring in issues of Bayesian approaches to probability and statistics. For readers interested in such matters we recommend the work of James Gleick (Gleick 1997). We shall explore Bayesian approaches in Sect. 3.6.

Fig. 3.1 Plotting the
Logistic equation. The
attractor plot for the
Feigenbaum Map with λ >
0.8 from Eidson et al. (1986).
Note that λ is k in our main
text notation—copyright
APS under licence RNP/24/
NO V/085254. All rights
reserved

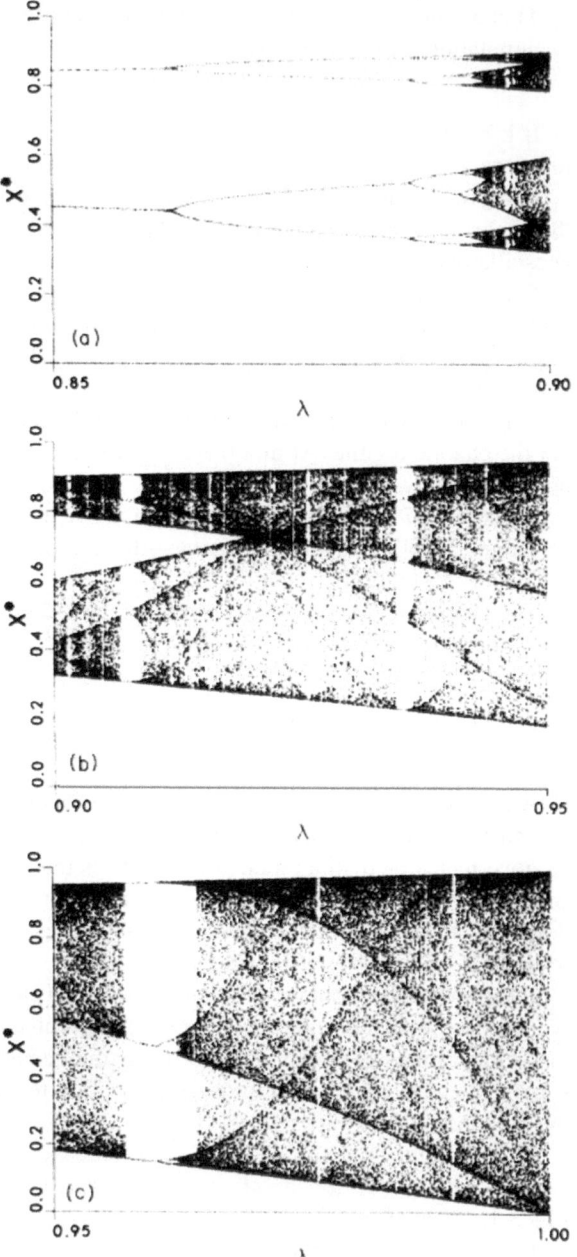

As noted earlier, the logistic function has its origins in an understanding of ecology and population dynamics. It was identified and explored by the Belgian mathematician Pierre François Verhulst (1804–1849), and was helpful in modelling population growth up to a system maximum (or carrying capacity). It was only 150 years later that the links to chaos were investigated in detail. Professor Robert May revealed and explained the realities in the mid-1970s (Stadlmann 2024).

To be clear, we are not saying that uncertainty in nuclear energy systems is governed by considerations of chaos in the formal mathematical sense. Our reference here to chaos is simply to make the point that even in deterministic systems there can still be genuine uncertainty.

The realities of chaotic systems remind us that we cannot simply rely on deterministic modelling. Uncertainty is something that is always present and should be explicitly considered by any competent engineer. Generally, this is a well understood concept, albeit perhaps implicitly. For example, engineers should know not to take the results of a computer simulation at face value as they may not exactly match real-world performance (due to uncertainty). Further, engineers are taught to consider tolerances in machining to ensure good fit of components as it is impossible for a machine or machinist to manufacture components of exact sizes consistently (uncertainty). However, we can go much deeper and create highly detailed analyses by quantifying the uncertainty and using it in the models that drive our decision making.

3.1.3 The Impossibility of Certainty?

The domain of mathematics known as Game Theory looks at games, including well established games played for amusement, with a view to seeing core mathematical realities. One consequential idea is the concept of a 'solved game'. Perhaps the most widely known solved game is noughts and crosses, or tic-tac-toe as it is known in America. Put simply the first player can be sure that they will not lose by choosing a corner square and then following a simple strategy, whilst the second player cannot win, whatever they do. The list of solved games is growing all the time, and it includes the popular pursuit 'Connect 4'—as sold by Hasbro under their copyright. Here we talk about a concept known as 'strongly solved'—for those wishing to know more we recommend the work of van den Herik and colleagues (van den Herik et al. 2002). Weaker solution states exist and there is much interest in whether draughts (checkers) or even chess might be solved with time and ever more powerful computers. The key to these ideas is determinism—the games referred to above are not like backgammon or card games played with some cards still in the deck—there is no randomness involved.

Now let us make a logical jump from games to sports. At some level the context remains deterministic, but the system admits aspects that are essentially impossible to characterise or describe.

For example, in the UK there is a popular table game derived from billiards called snooker. The best players are professional and indeed are famous celebrities. Snooker has the attribute that the starting player can achieve a perfect score, denying the other player any role in the game. This achievement is known as a "maximum break" or "147 break" referring to the maximum score. The first ever recorded maximum break was in 1955 achieved by Joe Davis (Guinness World Records 2024). The frequency has increased over the decades such that in a given professional competitive season a dozen or more maximum breaks might occur. While the impressive achievement is still part of the fun of the game, there lies in this reality a threat to the viability of the game as a whole. If a maximum break becomes routine then what is the point of a competitive sport where only the first player to play has a chance of winning or, thinking back to noughts and crossses, the game always ends in a stalemate?

Why is this happening? Is it that the equipment of the game is improving to avoid uncertainty? Are the players managing better to control uncertainty in their own performance. Snooker rather implies that irreducible uncertainty might be reduced to the point where it does not matter. Other sports with growing issues of a 'perfect score' problem include ten-pin bowling and darts. These three sports (snooker, darts and 10 pin bowling) share a feature. When a player is performing, their opponent is, temporarily, a spectator. Thus, the player is "competing" against the laws of physics. Most sports—basketball, football, cricket or baseball will never be solved games. The opposition is playing at the same time. The uncertainty inherent in not knowing what the opposition will do makes simply applying the laws of physics to predict the outcome impossible.

One last comment on determinism and games. Chess is a fiendishly complicated deterministic game. For the last 70 years computers have been playing chess with ever improving results (Wall 2024). Even without machine learning and artificial intelligence it is interesting to note that, by essentially brute force means, chess computers have been able to find winning strategies that thousands of years of human chess playing never found. The ability of technology to, in essence, *solve the game* is powerful and profound and it lies behind much of the thinking in this book.

Putting sports and games to one side, at this point we suggest it is helpful to pause and to consider the philosophical basis of uncertainty. While uncertainty and probabilistic thinking is ubiquitous in science and engineering, that familiarity masks some well-known subtleties. Some of the considerations are at the level of research philosophy, but that does not mean that they are not important in real engineering situations.

As the discussion of game theory reveals modern information technology is reducing the scope of uncertainty in ways that would have been unimaginable only 50 years ago.

3.2   Uncertainty in Practice

At this point it is important to note that uncertainty can be classified into two main categories: epistemic uncertainty and aleatory uncertainty. The former refers to things that we could in theory know better, but for practicalities, we do not. It represents the uncertainty implied by a lack of knowledge about a process, meaning that it is something that can be reduced or potentially eliminated given some effort to better understand it. Epistemic Uncertainty involves characterizing uncertainties using generalized probability (i.e., imprecise probability, fuzzy systems, Dempster–Shafer theory etc.) and seeks to narrow down uncertainty ranges by updating models as new information becomes available. It plays a crucial role in refining prediction ranges and making informed decisions by continuously improving our understanding of uncertain factors.

Aleatory uncertainty is the uncontrollable uncertainty, which is intrinsically part of what is under analysis, which cannot realistically be reduced by improved system knowledge.

To give an intuitive example, consider a weather prediction scenario where we want to determine if it will rain tomorrow. The weather is influenced by numerous random and arguably chaotic factors such as wind patterns, temperature fluctuations and moisture levels within the atmosphere. Even with the most advanced weather models, with the latest data, there is an element of 'chance' involved with any prediction. For example, chaos-based unpredictability is a form of aleatory uncertainty (see Sect. 3.1.2 for more on chaos). If, however, our weather model were creating predictions based on outdated data or missing key variables, such as changes in sea surface temperatures, then this lack of up-to-date understanding would be introducing epistemic uncertainty into our predictions.

In engineering applications, it is important to keep aleatory and epistemic uncertainty separated. By doing so, we can evaluate the potential improvement to the analyses of the system that can be achieved by reducing the epistemic uncertainty, or decide not to invest extra resources on a better estimation of some parameters due to their negligible effects on system performance and safety (a classic cost–benefit analysis). Mixing aleatory and epistemic uncertainty can be dangerous, as it provides a way to dilute uncertainty. If decision-makers do not clearly differentiate between what is fundamentally random (aleatory) and what is unknown due to lack of knowledge (epistemic), they may not allocate resources effectively to reduce uncertainty where possible. When both types of uncertainties are combined without proper distinction, the compounded effects can create scenarios where neither type of uncertainty is accurately represented.

In the world of engineering, we are already, perhaps subconsciously, aware of uncertainty. Any competent engineer dealing with manufacturing is aware of tolerancing—the allowable "wiggle room" within component dimensions that still allows for the desired fit and finish of an assembly. Such tolerances are created because we intuitively understand that machining comes with *aleatory* uncertainty—it's quite unlikely that a machine, even with the same machinist, can manufacture 1,000 components with each dimension the same down to the micrometre, there is an inherent unpredictability as to the resultant dimensions.[1] Thus, we effectively bound the dimensions with an interval uncertainty (this, and other types of uncertainty are

---

[1] For further information concerning the relationship between engineering tolerance and the safety margin, see the work of Brahma and colleagues (Brahma et al. 2024).

explained later in this chapter) to work with an understanding that components manufactured will have to fit with this range of dimension. We may be able to reduce some
of the epistemic uncertainty involved in this process, perhaps with more accurate
measurement tools or machines in the manufacturing process, but there will always
be an element of aleatory uncertainty. Of course, even with automated processes
there is still an issue of safe boundedness, but through automation it becomes easier
to reproduce physical realities more accurately and safe boundaries can be brought
tighter.

Former US Defence Secretary Donald Rumsfeld famous remarks regarding
"unknown unknowns" (USINFO 2002) have become a popular way to distil what
is, for most, a subtle and somewhat confusing idea. He stated that there are three
types of knowledge in decision making: "known knowns"—things we know that we
know, "known unknowns"—things that we know we do not know, and "unknown
unknowns"—things we don't know we don't know. This last category is perhaps the
most challenging in engineering project design—how can we prepare for something
we don't even know could happen? The story of the de Havilland Comet jet-powered
passenger aircraft of the 1950s is illustrative. It is presented in a text box.

> **How an Unknown Unknown Brought Down Britain's Foray into the Jet
> Age**
> The de Havilland Comet has cemented its place within aerospace, and engi-
> neering, history for both the right and the wrong reasons. It represented the
> best of British engineering, kickstarting the jet set age as aircraft moved from
> propeller-based propulsion to efficient and high-speed jet engines. Passengers
> would soar along at 750 km/h, rather than the prior 500 km/h, reaching their
> destinations faster while cruising higher above bad weather. The Comet's illus-
> trious story was significantly marred, however, by a series of infamous fatal
> accidents (Aviation Safety Network 2024). Three within a one-year period
> between 1953 and 1954 highlighted the impact of unknown unknowns in
> engineering design.
>
> The Comet was an advanced design featuring a fully pressurized cabin—
> a concept initially deployed by the British aircraft industry in the Bristol
> Brabazon of the late 1940s, an aircraft which had been a commercial failure.
> The Comet was even more radical in combining high-altitude cruising with jet
> engines that promised faster and smoother flights. The Comet had a sleek
> modern design that symbolized the future of air travel. During flight, the
> now higher cruising level means the fuselage of the aircraft experiences more
> vigorous pressurization and depressurization cycles every flight. These cycles
> induce stress on the aircraft's structure, particularly around openings such as
> windows and doors. The Comet's designers gave excited passengers superb
> views through relatively large rectangular windows (Fig. 3.2).

Fig. 3.2 Comet 1A Aircraft, with the rectangular passenger windows 1952. *Source* Wikimedia under GNU Free Documentation Licence—derived from work by Clinton Grove and edited by Altair78

The three catastrophic accidents occurred within a short span, each involving the disintegration of the aircraft mid-flight. Initially a frightening mystery, the root cause of these accidents was eventually traced to the Comet's square-cornered windows which exacerbated stress concentrations, and where riveting during manufacture could cause micro-cracks within the fuselage's skin which could grow into substantial cracks with the fatiguing of the structure. The public inquiry concluded that metal fatigue was the cause of the in-flight structural break-ups (BBC 2008). But why wasn't this noted during development and testing?

British Civil Airworthiness Requirements in 1949 required that fuselages have a proof pressure of 4/3 cabin pressure at attitude, and the design pressure be 2 times the cabin pressure at attitude. De Havilland used a higher rating of 2 times and 2.5 times, respectively. Overdesigning was an acceptable methodology at the time and the belief was that the higher proof stress would protect against fatigue. During development tests, cabin sections would be proof tested up to twice the operating pressure two times, and then cycled repeatedly up to the operating pressure. A test fuselage failed after 16,000 (simulated) flights, surpassing the legal requirement of 15,000 cycles. So then, why did two of the accident aircraft only perform 1290 and 900 flight cycles before they crashed (Davies and Birtles 1999)? What caused such a big difference between de Havilland's development tests, and the real aircraft?

At the tip of a fatigue crack, there is a high stress concentration. So, when the cabin is pressurised, a load is applied—this causes the yield stress of the material to be exceeded in a small area surrounding the crack tip. As a result, a small zone around the crack tip is plasticly deformed. However, the surrounding material is still elastic, meaning that when the load is removed again (i.e., when the cabin is depressurised after landing), the elastic material tries to return to its original shape but some of this material will find a lump of plasticly deformed material blocking this return. This results in a compressive stress on the plastic zone, squeezing it together. As the tip of the crack will be in the centre of the plastic zone, the end-result is that the sides of the crack are forced together. This makes it harder to grow the crack, so any crack growth will be slowed down. The larger the load you apply, the larger the plastic zone, and the stronger this effect will be. Normally a large load will cause a large amount of crack growth. However, if you apply just one large load, and then lots of small ones, the plastic zone caused by the initial large load will slow down the crack growth during the subsequent cycles, and it will take the crack longer to grow. This is what happened during de Havilland's testing of the fuselage during development.

The test fuselage used for the fatigue test was first used to check whether the fuselage design was resistant to over-pressurisation—it was pumped up to twice the pressure difference expected to be encountered in service. de Havilland's engineers applied a large load on the structure right before the fatigue test, causing large plastic deformations. Consequently, during the fatigue test cracks that were present in the fuselage grew much more slowly than those in the aircraft used in service. The safe life calculated from the fatigue test was therefore much too long.

Ultimately it was unknown fatigue interactions within the aluminium fuselage that created an unreliable fatigue test, that made the engineers sign off the Comet's design as safe and be blind to the development of fatigue cracks around the Comet's windows, even when the engineers overdesigned specifically to cater for their current knowledge of fatigue. The de Havilland Comet's tragic accidents underscored a fundamental lesson in engineering: the presence of an unknown unknown.

The story of the de Havilland Comet serves as a poignant reminder of the complexities and uncertainties inherent in engineering. It illustrates that even with comprehensive knowledge and testing, there can still be elements beyond our anticipation. The concept of unknown unknowns remains a critical consideration in all fields of engineering, driving continuous innovation, testing, and improvement to safeguard against unforeseen failures. As is all too common in aviation, the tragedy of these accidents created an equally large stride to improve our understanding and safety of aircraft. Following these investigations, the first methods for relating fatigue crack growth rate to the instantaneous crack length and applied stress were published in 1961, with Paris and Erdogan proposing their famous method of predicting fatigue crack growth in 1963 (Paris and Erdogan 1963).

The de Havilland Comet is a fascinating piece of human history, and its legacy is felt to this day across all fields of engineering. If you're keen to learn more, we recommend searching for articles and talks given by Professor Paul

Withey who is extremely knowledgeable in the aircraft's accident investigations and the engineering underpinning them.

References

1. Aviation Safety Network. ASN safety database (2024) https://asn.flightsaf ety.org/database/databases.php. Accessed Date 31 Oct 2024

2. BBC (2008) On this day—19th October. http://news.bbc.co.uk/onthisday/ hi/dates/stories/october/19/newsid_3112000/3112466.stm. Accessed Date 31 Oct 2024

3. Davies R, Birtles PJ (1999) Comet: the world's first jet airliner: Paladwr Press, ISBN: 10: 1888962143

4. Paris P, Erdogan F (1963) A critical analysis of crack propagation laws. J Basic Eng 85:528–533. https://doi.org/10.1115/1.3656900

In current practice, unknowns in engineering systems are typically handled with safety factors (see Sect. 2.1) and the tendency to over-design. Safety factors are multipliers applied to design specifications to account for uncertainties and potential variations in material properties, environmental conditions, or unexpected loads. This practice helps ensure that even if unforeseen circumstances arise, the system will perform reliably and safely. Engineers often lean towards over-designing components, intentionally building systems to be more robust than the minimum required. This precautionary approach helps mitigate the risks associated with unknown unknowns, albeit with a lack of nuance and quantification of the uncertainty at play.

As a result of uncertainty, we cannot simply rely on deterministic modelling and uncertainty is something that is always present and should be explicitly considered by any competent engineer. Generally, uncertainty is a well understood concept, albeit perhaps implicitly, such as with manufacturing tolerances. However, we can go much deeper and create highly detailed analyses by quantifying and modelling the uncertainty at play and using it in the models that drive our decision making.

Engineering resilience is the ability of an engineered item, such as a structure, to withstand external disturbance such that its core functions are not impaired in the long-term. The concept links to a sub-discipline of engineering known as Resilience Engineering. Resilience Engineering forms part of safety engineering and it is about designing a system, typically a complex adaptive system, such that it can withstand external shocks. The concepts are not only applied to electromechanical systems— they can be applied to human organisations and even society itself. One can design for more resilient socio-technical systems via resilience engineering. Thinking in such terms, Erik Hollnagel describes the core attributes of resilience engineering as being based on how a system is designed with respect to four core considerations (Hollnagel 2024):

- how it responds
- how it monitors
- how it learns, and
- how it anticipates.

In this way resilience is more than just a passive response, as might be seen in an elastically stressed structural component of a building, or an aircraft (see text box). Once the external stress is removed the structure spontaneously returns to normal. Resilience engineering these-days usually involves active IT-supported intervention and restoration of state (reset). It remains to be seen what the full impact of generative artificial intelligence and machine learning will be on resilience engineering as a field. The potential for a significant boost in performance and efficiency seems to be great.

3.3 Multiple Dimensions of Uncertainty

New information technology capabilities developed over the last 40 years have allowed us to represent uncertainty in multidimensional heat maps—as shown in Fig. 3.3. These computational developments largely came after the maturing of nuclear power in the period 1960–1980. We shall return to these ideas later in Chap. 5.

As we here consider multiple dimensions of uncertainty, one must not forget that even within a single dimension of uncertainty there can be a source of imprecision in understanding and from that a risk of misunderstanding.

Dr Nawal Prinja has pointed to the high level of multi dimensionality of modern nuclear power safety-engineering. He points to the advanced materials science deployed in the service of nuclear fusion engineering, for example. Inspired by such thinking let us unpack just one aspect of fission nuclear technology relevant to power station performance—the nuclear fuel. In recent years there has been much interest in nuclear fuels with more complex micro-structures. Part of this thinking has been motivated by a desire to develop accident tolerant fuels. More complex fuel stoi-chiometries relate to the notion of 'high entropy alloys'. These may have multiple chemical elements involved, perhaps as many as four or five rather than simply metallic uranium or uranium dioxide as used historically. Such new fuels remind us that each chemical element arguably represents a dimension of the problem. In addition, such advanced materials usually require more sophisticated material processing techniques and hence in addition, each such technique is also arguably a further dimension of the engineering analysis.

The multidimensionality of uncertainty in advanced engineering systems is an important consideration. Conceptually it points to the notion of independent (orthogonal) uncertainty considerations. In reality, of course, many uncertainties have inter-dependencies, even potential causal relationships, but the notion of dimensionality remains important. Uncertainty sits within a multidimensional space which can be structured with reference to wholly independent (orthogonal) parameters. In some cases, these axis parameters will be clear measurable attributes of some aspect of the

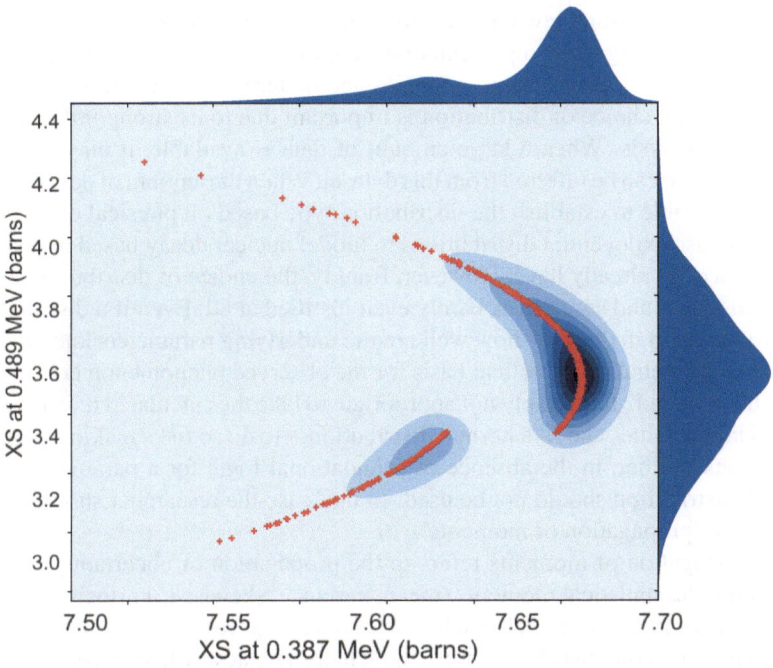

Fig. 3.3 Two dimensional uncertainty heat map showing projections in each dimension. This data relates to nuclear scattering cross-section measurements in Fe-56. Further details are given in the text box entitled Nuclear Data Sampling in Sect. 3.7

system. With all that said, we shall hold to the view that time is special, and it should not simply be regarded as one axis of the multidimensional uncertainty space.

3.4 Uncertainty Representations

3.4.1 Modelling Data with Distributions

Possibly the most informative way of quantifying uncertainty is by describing the data with a distribution, such as a Normal/Gaussian, Uniform, Beta or Gamma distribution. The probability density functions (PDFs) of these distributions describe the likelihood of a range of potential outcomes. Normal distributions are popular in finance to model the return of stocks and have statistics attached to them such as the mean and variance. Exponential or Weibull distributions are common in reliability analyses to model the expected time until failure of components in systems, enabling a probabilistic viewpoint towards maintenance schedules or safety assessments. By leveraging these representations, engineering decision-makers can make more informed choices, optimize strategies and better anticipate future events.

Naturally, the choice of distribution is important due to its strong influence upon uncertainty analysis. When a large amount of data is available, it may be that the distribution type can be inferred from this dataset. When the amount of data is limited, it may be possible to establish the distribution type based on physical expectations, i.e., we can use exponential distributions to model nuclear decay based upon empirical evidence we already have. However, frankly, the choice of distribution is often weakly justified, and sometimes barely even justified at all. Even if a distribution is correctly justified in theory—how well are the underlying parameters known? In the absence of any reliable theoretical basis for the observed phenomenon how should a researcher respond? It is simply not appropriate to take the calculated mean and standard deviation values and fit a normal distribution—to do so risks making dangerous assumptions. Rather, in the absence of foundational logic for a parameter's distribution, a distribution should not be used. In that case the researcher should think in terms of the 'propagation of moments'.

The propagation of moments refers to the propagation of uncertainty by purely propagating the statistical moments (mean, variance, skewness, kurtosis, etc.) of the uncertainty through the model. Such a procedure requires less data than would be required to infer a full distribution type and it helps negate any bias introduced by our assumptions of distribution type. Of course, moment propagation limits our resultant analyses to the moments, and hence the nuances of the probabilistic spread will be unavailable.

In Fig. 3.4 we illustrate longitudinal plane wave scatter in 2-D. Sound in air is an example of a longitudinal plane wave. Extending our thinking will take us to the world of light propagation and light scattering, which, like the scattering of sound waves, is more than just a two-dimensional problem.

In Fig. 3.5 we refer to an illustration of a heat map in three dimensions and in the Figure caption we refer to the idea that such a situation might be regarded as 'four dimensional'. In this book we would not regard this as being a 4D situation owing to our distinction between dependent and independent terms. The orthogonal axes permit the variation of independent terms in three dimensions (here represented on a two-dimensional page), while the heat map shows a dependent quantity.

Consideration of the relationship between computer graphics and data visualisation takes us in the 2020s to the mathematics of ray tracing and computer graphics. In Fig. 3.6 we present a wholly synthetic image constructed by a computer from minimal exogenous inputs and invoking only the laws of physics relating to optical phenomena. The level of naturalism obtained is quite remarkable.

Earlier in Sect. 3.1 we discussed the impossibility of certainty, albeit in respect of specific conditions (e.g. deterministic chaos or quantum phenomena). Thankfully these realities of fundamental indeterminacy are restricted to specific special contexts and hence in our daily experience are largely avoided. The power of deterministic physics in our everyday experience is revealed by Fig. 3.6. That Figure shows that an

Fig. 3.4 2D Modelling of longitudinal plane wave scatter: Copyright CC-BY-4.0. *Source* Sha and Albannai (2024)

Fig. 3.5 Two-dimensional representation of a heat map in three orthogonal dimensions. The reported heat map value is arguably a fourth dimension, but this is not a framing that we adopt here—see main text. *Image Source* Modave et al. (2020). Used under CCC Licence 5903660996313. All rights reserved

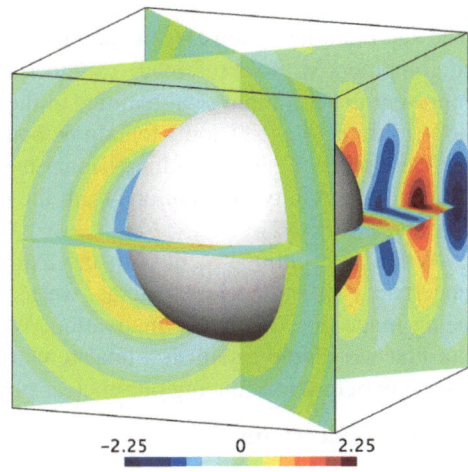

IT model based on deterministic optical physics can yield images essentially indistinguishable from a photograph of the natural world. That is the natural world, in this context can be persuasively represented on the basis of uncertainty free modelling. In some ways this is a visual metaphor for the Turing Test in Artificial Intelligence[2]

---

[2] The classical Turing Test was first described by British mathematician Alan Turing in 1949. Originally termed the imitation game, Turing posited the question: "Can machines think?". He proposed

Fig. 3.6 Synthetic still life image created in software using ray tracing. *Source* Giles Tran 2006 Wikimedia—Originator grants copyright release for any legal purpose

The key point here, however, is the fact that observable realities can be reproduced in computers without reference to uncertainty does not mean that computers can represent uncertain physical realities in the virtual space with certainty. All we can hope to do is to use IT tools to help us better understand uncertainty. For example, we can appreciate its multidimensional nature (see Sect. 3.3), and we can attempt to track uncertainty distributions from the component level up to the whole system level (see Sect. 2.2).

While we continue to hold to the axiom that there will always be irreducible uncertainty in engineered systems, we also acknowledge the enormous advances being made possible by IT innovation. We shall return to such ideas in Chap. 5, but first we should establish a final few attributes of uncertainty and some further concepts before introducing our nuclear power case study—the UK AGR.

Bounded Uncertainty

In the complex world of decision-making, uncertainty is a constant companion. Whether in business, science, or everyday life, we rarely have access to complete information or perfect foresight. Yet, in many situations, this uncertainty is not boundless—it comes with certain constraints or limits. This is where the concept of Bounded

---

a text-based test in which a subject would engage in a text mediated conversation with a person and a machine. The issue was whether the subject could detect the mechanical counterparty. Turing stressed that the issue would not be whether the machine could answer the questions accurately the test would be could a machine communicate like a human?

Uncertainty becomes essential. By understanding and quantifying these limits, we can make more informed and confident choices, even in the face of incomplete knowledge.

3.4.2 Bounded Values with Intervals

In mathematics, an interval refers to a range of numbers between two given endpoints. It represents all the numbers that lie between (and possibly including) these endpoints. Intervals are commonly used in real analysis and calculus to describe subsets of real numbers. For example, the interval [0, 7] includes all numbers greater than or equal to 0 and less than or equal to 7.

In quantifying uncertainty, an interval typically implies a large amount of uncertainty and effectively says "its true value is somewhere in this range" while providing no suggestion of which value it may frequently take (a mean) or its spread (a variance). These qualities make intervals a good choice for quantifying uncertainty in situations where there is a limited amount of data available on the spread of the variable, but we have some intuition towards its minimum and maximum values.

We, as people, are subconsciously bounding values with intervals in our day-to-day life. Humans may not be good at estimating the exact value of some phenomenon (or variable), but we are usually quite good at predicting a range, i.e., an interval. Engineers are used to working with intervals—tolerance in manufacturing is an example of an everyday use of intervals, with the client providing a plus/minus tolerance giving bounds on dimensions to ensure correct fitting of components after manufacture as there is an understanding that the machine and machinist come with their own uncertainty. Intervals can also come from physical constraints; liquid water is generally between 0 and 100 °C. Any measurement which is taken by an instrument with an associated accuracy can and should be represented by an interval.

Even when we use distributions to characterise a variable affected by uncertainty, we use confidence intervals to assign a level of credibility to an interval. For instance, estimating the parameter of a binomial distribution should involve calculating a confidence interval around the estimated proportion. This interval provides a range within which we expect the true proportion to lie, given a certain level of confidence (commonly 95% or 99%). By using confidence intervals, we acknowledge, and account for, the inherent variability and uncertainty in our estimates. This approach not only provides a point estimate, but it also quantifies the uncertainty associated with it, allowing for more informed decision-making and a better understanding of the possible variability in the data. Confidence intervals serve as a critical tool in statistical inference, enabling us to make probabilistic statements about population parameters based on sample data.

The propagation of intervals through engineering models (see Fig. 3.7) introduces computational challenges even today. In fact, even a simple interval arithmetic operation such as the summation of n intervals requires 2^n operations. If the problem is not monotonic the propagation of the interval though a model requires solving

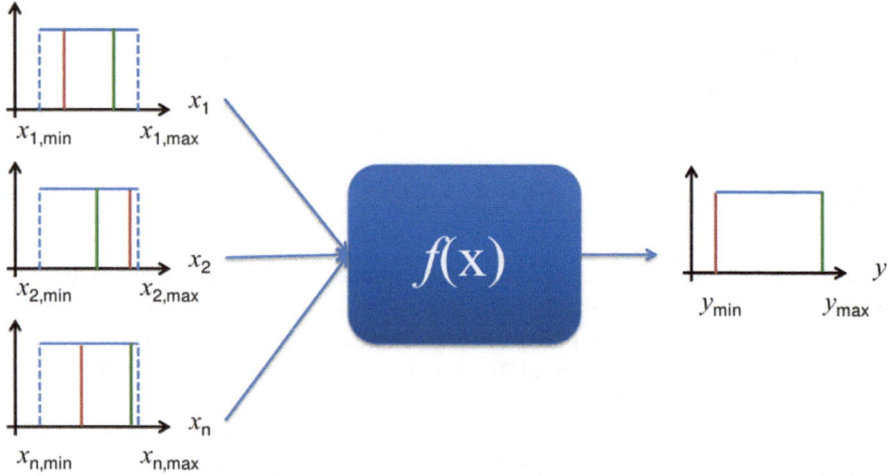

Fig. 3.7 An example interval propagation. The three parameters on the left are constrained with intervals (vertical dashed lines). Naturally, the combination of input parameters that maximizes (green lines) or minimizes (red lines) the function output may not lie at each parameter's bounds and must be found with a bounded optimization process. *Source* used with kind permission of Edoardo Patelli.

two optimisation problems to find the minimum and maximum of the response, i.e. identify the values of the input parameters that produce the extreme values of the response. Depending on the complexity of one's model and the chosen optimization algorithm, this can become computationally exhausting.

3.4.3 Combining Uncertainty with Probability-Boxes (P-Boxes)

When we may have a preconceived notion on a dataset's distribution model (e.g., Normal, Beta, Uniform, etc.) but remain uncertain as to that model's parameters it doesn't make sense to use deterministic values for these parameters as we are creating an inaccurate model of our present understanding. In such cases, we can bound the parameters with interval uncertainty to represent our uncertainty on the distribution parameters. This scenario can consider both epistemic and aleatory uncertainty (see Sect. 3.2), because the distribution models the natural, irreducible variation in the data (aleatory) while the intervalised parameters represent our lack of complete knowledge (epistemic). Such an approach is called a probability box (P-box) as it creates minimum and maximum bounds upon the cumulative distribution function (CDF) of the distribution model (see Fig. 3.8). The concept of the P-box dates back to the pioneering work of Scott Ferson and colleagues (Ferson and Ginzburg 1996).

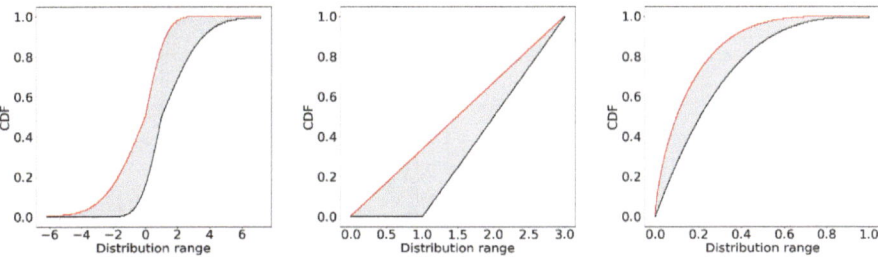

Fig. 3.8 From left to right, Normal distribution with mean in the range [0,1] and variance in the range [1,2], Uniform distribution with left bound equal to [0, 1] and right bound equal to 3, and Beta distribution with alpha in the range [0.7,1] and beta in the range [1, 2]. *Source* used with kind permission of Ewan Smith. All rights reserved

Without the use of more advanced propagation approaches, the direct, Monte Carlo propagation of P-boxes is possible, but it is a computationally intensive endeavour. A common approach is double-loop sampling, where an N number of random "alpha cuts" are taken of parameter CDFs between 0 and 1, with an inner optimization loop bounded by the CDF interval used to determine the minimum and maximum of the model output given the parameter interval. Readers may find it useful to note that advanced propagation methods do exist. See for example: (Faes et al. 2021).

3.4.4 Reachability Analysis

Reachability analysis is a method used in systems theory and control to determine the set of states a dynamic system (systems modelled by ordinary differential equations where the system evolves over time) can reach, starting from an initial condition and following its governing equations over time. It is a crucial tool for verifying the safety and performance of both deterministic and non-deterministic systems. By analysing reachable sets, engineers can predict whether a system will stay within safe operational limits, avoid obstacles, or meet other predefined criteria, helping ensure robustness and reliability in working environments.

Analyses are conducted by defining a possible set of initial conditions/initial states and propagating them through to the desired time horizon. Outside the world of simple models, an exact solution of complex and/or black-box models is rarely computable, so a discrete outer approximation is usually calculated (such as by bounding the set with intervals). Figure 3.9 captures both input and discretisation uncertainty. These outer approximations then provide a quantification of the system's 'reach' of possible states given the potential range of initial system states, Fig. 3.10.

A failure region can be defined within the analysis—it is a system threshold or combination of system states which leads to failure or unsafe operation of the system under analysis, e.g. thermal meltdown, current draw greater than capacity, etc. (see Fig. 3.11).

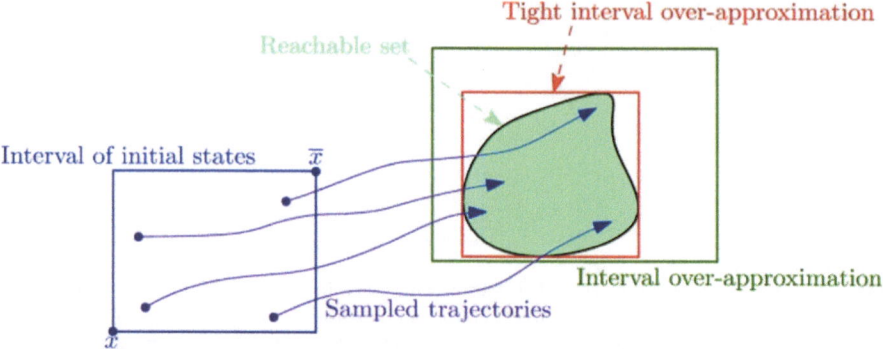

Fig. 3.9 Using intervals to over-approximate reachability analyses when an exact solution is hard to compute. *Source* Meyer et al. (2021). Used under CCC Licence 5903690742904. All rights reserved

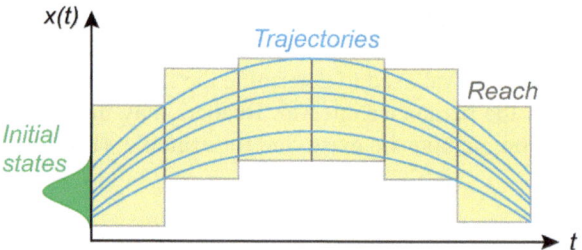

Fig. 3.10 Diagram of reachability analysis. Initial system state is uncertain and described with a distribution—several trajectories are sampled from the initial state and propagated along the time horizon. *Source* used with kind permission of Ewan Smith. All rights reserved

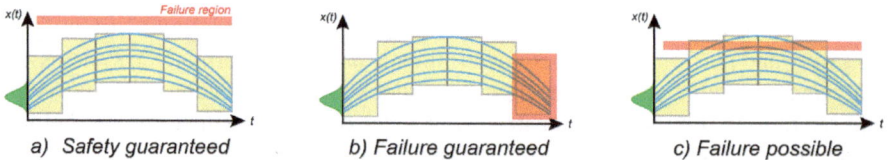

Fig. 3.11 Three example reachability scenarios when considering a failure region in reachability analysis. **a** When the failure region is completely out with the reach of the analysis, system safety is guaranteed. **b** Should the failure region completely engulf a portion the reach, failure is guaranteed around this point in the time domain. **c** When the failure region only occupies a portion of the reach, failure is possible, i.e., the probability of failure lies between 0 and 1. *Source* used with kind permission of Ewan Smith. All rights reserved

3.5   Uncertainty Sensitivity

Fundamentally, sensitivity refers to how a system, or model, responds to changes in input parameters. It's about how much a small change in an input will affect the output. In different contexts, this can mean physical systems, mathematical models, or simulations. For example, imagine you are controlling a car. Sensitivity in this case would describe how much the car turns when you move the steering wheel. If the car turns sharply with a small movement of the wheel, it's highly sensitive. If it only turns slightly, it's less sensitive.

In engineering, sensitivity is a measure of how much the output of a system, or a model responds to slight variations in its input parameters. In control systems, for example, it reflects how well the system can tolerate disturbances or parameter variations while maintaining stable performance. Sensitivity analysis helps engineers identify which variables have the most significant influence on the system's behaviour, allowing for better understanding and the optimization of designs for robustness. By systematically adjusting inputs and observing the corresponding changes in outputs, sensitivity analysis can reveal potential weaknesses and guide resource allocation in the design process. This method is particularly valuable in complex systems where multiple interacting factors contribute to performance, making it an essential aspect of model validation and risk assessment in engineering.

Sensitivity analyses offer several benefits. They can:

- Determine the robustness of a model to variations in inputs

  - When validating a model, one needs to understand its sensitivity to uncertainties. If slight changes in input variables cause large deviations in results, the model may not be reliable for predicting real-world behaviour, signalling a need for further refinement. Sensitivity analysis helps in determining which parameters should be carefully controlled and validated during experiments.

- Assess the impact of assumptions and simplifications made during the modelling process
- Provide insight into how the system behaves in response to input value variation and can determine the overall impact of each input's uncertainty upon the model output
- Gain an understanding of the interaction between multiple inputs
- Reduce model dimensionality by removing insensitive parameters.

A multitude of sensitivity analysis approaches exist in literature; however, most adhere to a typical set of general steps:

1. Run the model a number of times by perturbing the input parameter around a specific point (local sensitivity)
2. Run the model a number of times using some sort of design of experiment (global sensitivity)
3. Using the obtained outputs, calculate the sensitivity measure of interest.

Sensitivity measures are quantitative metrics that describe how variations in input parameters influence the output of a model. Some commonly used sensitivity measures include Sobol indices (Sobol', 2001), which partition the variance in the output among the different inputs to indicate their contribution to total output variability, the Partial Rank Correlation Coefficient (PRCC) (National Institute of S&T 2024) or the Morris method (Saltelli et al. 2002). These provide insight into both the magnitude and direction of each input's influence. These measures are vital for identifying key drivers of uncertainty and for pinpointing which parameters should be prioritized when allocating resources towards uncertainty reduction/data collection. They also help determine non-influential parameters, allowing for model simplification by reducing the number of variables without compromising the accuracy of predictions. By employing these sensitivity measures, engineers can better understand complex, non-linear interactions and enhance the robustness and reliability of their models.

3.6 Bayesian Statistics

If you spend a reasonable amount of time around a statistician, or with your head in a statistics textbook, you will probably come across mention of Bayesian statistics—what is this idea?

Bayesian statistics is a powerful approach to statistical inference that uses probability to quantify uncertainty in decision-making. Named after the 18th-century mathematician Thomas Bayes, it provides a formal method for updating the probability of a hypothesis as new evidence or information becomes available. Unlike traditional frequentist statistics, which focuses on long-run frequencies of events, Bayesian statistics incorporates prior beliefs or knowledge into the analysis, allowing for a more flexible and intuitive framework. This approach is widely used in fields where decision-making under uncertainty is crucial.

At the core is Bayes' Theorem—a way to reverse conditional probabilities. While conditional probability tells us the likelihood of an event given that another event has occurred, Bayes' Theorem helps us answer the question: "Given that a certain outcome has occurred, how likely was a related event the cause of this outcome?".

The mathematical form of Bayes' Theorem follows:

$$P(A|B) = \frac{P(B|A) \cdot P(A)}{P(B)}$$

where $P(A|B)$ is the posterior probability, the probability of event A occurring given that B is true, $P(B|A)$ is the likelihood, the probability of event B given that A is true, $P(A)$ is the prior probability of event A, and $P(B)$ is the marginal likelihood, or the total probability of B occurring regardless of A. In simpler terms, Bayes' Theorem helps us revise our initial estimate (the prior) of how likely something is, based on the discovery of new information.

Let's imagine a medical scenario where a patient is being tested for a rare disease. Only 1% of the population has this disease, and a certain test for it is 90% accurate. That means if a person has the disease, the test will correctly identify them 90% of the time. However, the test also has a 5% false positive rate; meaning 5% of people without the disease will still test positive. Now, suppose the patient tests positive. What is the probability that they actually have the disease? We can apply Bayes' Theorem to calculate the probability that the patient has the disease given that they tested positive:

$$P(Disease|Positive) = \frac{P(Positive|Disease) \cdot P(Disease)}{P(Positive)}$$

To find $P(Positive)$, the total probability of testing positive, we need to consider both the true positives and the false positives:

$$P(Positive) = P(Positive|Disease) \cdot P(Disease)$$
$$+ P(Positive|NoDisease) \cdot P(NoDisease)$$
$$P(Positive) = (0.9 \cdot 0.01) + (0.05 \cdot 0.99) = 0.0585$$

Now, we can plug this back into Bayes' Theorem:

$$P(Disease|Positive) = \frac{0.9 \cdot 0.01}{0.0585} = 0.1538$$

Hence, one can conclude that there is a roughly 15% chance that our patient actually has the disease.

Bayes' Theorem shows how new evidence (a positive test) doesn't necessarily mean the outcome (having the disease) is as certain as it might seem at first glance. This theorem is widely used in various fields due to its power to enhance predictions by considering both prior beliefs and new data.

For more on Bayesian statistics, we recommend: Johnson et al. (Johnson et al. 2002).

Nuclear Data Sampling—Getting Technical

In the nuclear power sector, it is important to have a firm understanding of nuclear data. Indeed, such considerations formed a specific and separate part of the eMEANSS project. Further details are available on the eMEANSS website (www.nubu.nu/emeanss).

How is such data accommodated and interpreted? In essence there are two approaches. One approach uses covariance and assumes normality: This method involves generating a correlated Gaussian random vector, often using techniques such as Cholesky factorization of the covariance matrix. This approach can also be applied to continuous functions, such as a cross-section, and is similar to methods used in Gaussian process regression and Gaussian random fields. In the context of nuclear data, this method is available through SANDY (Fiorito 2017). SANDY is a python package that can read, write and perform a set of operations on nuclear data files in the Evaluated Nuclear Data File (ENDF-6) format.

The second approach is a Bayesian method with a nuclear reaction model code: This approach uses a nuclear reaction model code, such as TALYS, to produce a complete evaluated dataset. Bayesian updating is employed to build distributions for the input parameters of TALYS (Koning et al 2012) using available experimental data. The input distributions are then propagated through TALYS (Koning et al. 2012) using Monte Carlo simulations to create a distribution of evaluated nuclear data quantities. Unlike the covariance method, the resulting distribution from Bayesian updating is typically non-Gaussian (Fiorito et al. 2017). An example was illustrated earlier in this book in Fig. 3.3, which shows a TALYS Evaluated Nuclear Data Library (TENDL) scattering cross-section of Fe-56, with cross-sections at 0.387 MeV on the x-axis and at 0.489 MeV on the y-axis.

For Monte Carlo particle transport applications, this uncertainty propagation scheme can be viewed as second-order Monte Carlo. Each Monte Carlo particle transport simulation run generates a distribution (with a mean value and deviation corresponding to Monte Carlo error), and repeated executions produce a set of such distributions. The envelope of these distributions provides all the possible outcomes, owing to the epistemic uncertainty in the nuclear data.

References

1. Fiorito L, Žerovnik G, Stankovskiy A, Van den Eynde G, Labeau PE (2017) Nuclear data uncertainty propagation to integral responses using SANDY. Ann Nucl Energy 101: 359–366. https://doi.org/10.1016/j.anucene.2016.11.026

2. Koning A.J, Rochman D (2012) Modern nuclear data evaluation with the TALYS Code System. Nucl Data Sheets 113(12):2841–2934. https://doi.org/10.1016/j.Nds.2012.11.002

3.7 Uncertainty Updating

This chapter has repeatedly returned to the importance of epistemic uncertainty and its definition as a reducible uncertainty—but what does an uncertainty reduction process actually look like? How can we incorporate new data/measurements to update our uncertainty models?

Imagine you're doing a science experiment, like predicting how fast a ball will roll down a hill. You use a simple equation, but when you test it, the ball doesn't roll quite as fast as you predicted. Model updating is when you go back and change your prediction method (the model) to match what you actually observed in the experiment. Maybe you forgot to include the effect of friction, or you didn't account for air resistance properly. Once you adjust your equation (model) to match the real-world result, that's updating the model. Model updating means adjusting your prediction method so it better matches real-world results. As another example, in structural dynamics, you might develop a finite element (FE) model of a bridge, but after testing the actual structure, you find that the model's predictions for natural frequencies are slightly off. To improve the accuracy, you perform model updating by adjusting properties like material stiffness, mass distribution, or damping coefficients in the FE model until the predictions closely align with the measured data. In the literature, the updating of model parameters with measurement data is known as "model updating" (see text box).

Model Updating—The Technicalities

In the deterministic domain, model updating essentially distils down to an optimization process in which model parameters are adjusted to minimize the difference between model outputs and obtained measurements. In the probabilistic sense, the aim is for the distribution of parameters to propagate into a model output distribution which matches a distribution of repeated measurements, noting some measurement error variance. The vast majority of the probabilistic model updating literature has its underpinnings within Bayesian updating, the process of updating probabilities with Bayes' theorem:

$$P(\theta|y) = \frac{P(y|\theta) \cdot P(\theta)}{P(y)}$$

where θ is a vector of uncertain parameters, y is a vector of measurement data, $P(\theta|y)$ is the posterior (updated probabilities), $P(y|\theta)$ is the likelihood (the likelihood of seeing the measurement data given the parameters), $P(\theta)$ is the prior distribution (our previous belief) and $P(y)$ is the probability of seeing the measurements (a normalizing factor which ensures the probabilities add up to 1) (Fig. 3.12).

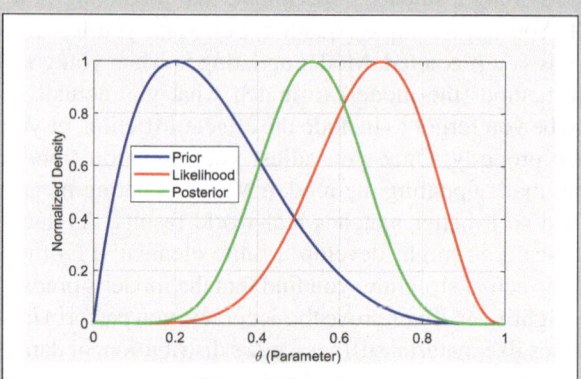

Fig. 3.12 Example probability distributions of the prior, posterior and likelihood within Bayes' Theorem. *Source* used with kind permission of Ewan Smith. All rights reserved

It is generally appropriate to assume that measured uncertainties follow a Gaussian distribution, and as a consequence it is relatively straightforward to define a likelihood:

$$P(y|\theta) = \left(\prod_{k=1}^{n} \frac{1}{\sigma_k \cdot \sqrt{2\pi}}\right) \cdot \exp\left(-\sum_{k=1}^{n} \frac{(y_k - f(\theta))^2}{2\sigma_k^2}\right)$$

where y_k is the kth measurement, σ_k is the measurement variance for the kth measurement and $f(\theta)$ is the model output using the given parameters.

However, it is not always suitable to assume a Gaussian uncertainty distribution, and as such different likelihoods based upon different distributions may be required. With a given likelihood definition in place, and noting the insights of Thomas Bayes, the posterior distribution can in principle be sampled from via a sampling algorithm such as Markov-Chain Monte Carlo (MCMC).

Further Reading

1. Lye A, Cicirello A, Patelli E (2021) Sampling methods for solving Bayesian model updating problems: a tutorial.Mech Syst Signal Pr 159:107760. https://doi.org/10.1016/j.ymssp.2021.107760

2. van Ravenzwaaij D, Cassey P, Brown SD (2018) A simple introduction to Markov Chain Monte–Carlo sampling. Psychon Bull Rev 25(1):143–154. https://doi.org/10.3758/s13423-016-1015-8

Acknowledgement This work was supported by the Engineering and Physical Sciences Research Council, UK via a grant entitled: *Enhanced Methodologies for Advanced Nuclear System Safety (eMEANSS)*. The grant had reference: EP/T016329/1. The authors thank the EPSRC for this support.

References

Brahma A, Ferguson S, Eckert C, Isaksson O (2024) Margins in design—review of related concepts and methods. J Eng Des 35. https://doi.org/10.1080/09544828.2023.2225842

Eidson J, Flynn S, Holm C, Weeks D, Fox RF (1986) Elementary explanation of boundary shading in chaotic-attractor plots for the Feigenbaum map and the circle map. Phys Rev A 33:2809–2812. https://doi.org/10.1103/PhysRevA.33.2809

Faes MGR, Daub M, Marelli S, Patelli E, Beer M (2021) Engineering analysis with probability boxes: a review on computational methods. Struct Saf 93:102092. https://doi.org/10.1016/j.strusafe.2021.102092

Ferson S, Ginzburg SR (1996) Different methods are needed to propagate ignorance and variability. Reliab Eng Syst Saf 54:133–144. https://doi.org/10.1016/S0951-8320(96)00071-3

Gleick J (1997) Chaos - making a new science, Vintage. ISBN: 0-7493-8606-1

Gray A, Davis A, Patelli E (2021) Uncertainty propagation in SINBAD fusion benchmarks with total Monte Carlo and imprecise probabilities. Fusion Sci Technol 77(7–8):802–812. https://doi.org/10.1080/15361055.2021.1895667

Guinness World Records (2024) First official 147 break in snooker. https://www.guinnessworldrecords.com/world-records/first-official-147-break-in-snooker/ Accessed Date 28 Sept 2024

Heisenberg W (1927) Über den anschaulichen Inhalt der quantentheoretischen Kinematik und Mechanik. Z Phys 43(3):172–198. https://doi.org/10.1007/BF01397280

Hollnagel E (2024) Resilience engineering. https://erikhollnagel.com/ideas/resilience-engineering.html. Accessed Date 16 Sept 2024

Johnson AA, Ott MQ, Dogucu M (2002) Bayes Rules!: an introduction to applied Bayesian modeling. Routledge. ISBN: 9780367255398

Meyer PJ, Devonport A, Arcak M (2021) Interval reachability analysis: bounding trajectories of uncertain systems with boxes for control and verification. Springer. ISBN 978-3-030-65109-1

Modave A, Geuzaine, A, Antoine X (2020) Corner treatments for high-order local absorbing boundary conditions in high-frequency acoustic scattering. J Comput Phys 401:109029. https://orbi.uliege.be/bitstream/2268/259058/1/2020_modave_corner.pdf

National Institute of Science and Technology (2024) Partial rank correlation. https://www.itl.nist.gov/div898/software/dataplot/refman2/auxillar/partraco.htm Accessed Date 20 Nov 2024

Saltelli A, Tarantola S, Campolongo F, Ratto M (2002) Sensitivity analysis in practice: a guide to assessing scientific models. Wiley. ISBN: 9780470870938

Sha G, Albannai A (2024) Recent progress in ultrasonic nondestructive characterization of material microstructures: a review. J Nondestruct Test https://www.ndt.net/article/ndtnet/papers/Recent_Progress_in_Ultrasonic_Nondestructive_Characterization_of_Metal_Microstructures_Review.pdf

Sobol' IM (2001) Global sensitivity indices for nonlinear mathematical models and their Monte Carlo estimates. Math Comput Simul 55(1–3):271–280. ISSN 0378-4754. https://doi.org/10.1016/S0378-4754(00)00270-6

Stadlmann J (2024) https://www.merton.ox.ac.uk/sites/default/files/inline-files/Robert-May.pdf. Accessed Date 20 Sept 2024

USINFO (2002) Defense department briefing February 12, 2002. https://usinfo.org/wf-archive/2002/020212/epf202.htm Accessed Date 23 July 2024

van den Herik HJ, Uiterwijk JWHM, van Rijswijck J (2002) Games solved: now and in the future. Artif Intel 134:1–2. https://doi.org/10.1016/S0004-3702(01)00152-7

Wall B (2024) Early computer chess programs. https://archive.ph/20120721202324/http://www.chessville.com/BillWall/EarlyComputerChessPrograms.htm. Accessed Date 28 Sept 2024

Williams GP (1997) Chaos theory tamed. Henry (Joseph) Press. ISBN: 0-7484-0749-9

Chapter 4
Nuclear Power—Our Reference Technology

4.1 Introduction

The UK Advanced Gas-cooled Reactor (AGR) design is used in this book as the basis of examples of the uncertainty propagation considerations outlined in the previous chapters. The AGR was chosen as the basis for these studies as it is a well understood and mature design, and it provides a good basis for the examples of the application of the new approach to risk. While the UK AGR is not a contender in the competition around 21st Century nuclear renaissance, it has inspired some forward-looking high temperature reactor ideas - in particular, we note: (Margulis and Shwageraus 2021). The two case studies illustrate the use of the approach in a scenario developing over a long period of time (Sect. 5.2) and one occurring over a short time scale (Sect. 5.3). The methods described are applicable to other reactor designs and to systems outside the nuclear industry, although the specifics differ, of course.

The AGR design followed on from the first UK commercial reactor design, the Magnox reactors. Both designs were gas cooled and use graphite as the moderator. The AGR programme was beset by problems in the early stages with cost and time overruns. Construction of the first commercial[1] AGR power plant commenced at Dungeness in 1965 and the station was finally connected to the electricity grid in April 1983.[2] The final AGR power station started construction in 1980 and was connected to the grid in 1998.

At the time of writing, in late 2024, AGRs comprise 8 of the 9 commercial reactors[3] operating in the UK (ONR 2024a) and some of these are expected to remain in operation until the late 2020s. Lifetime extensions beyond this date are being considered for some of these (Mavrokefalidis 2024).

---

[1] The Windscale AGR (WAGR) was a prototype plant.

[2] Due to this extended construction time the first AGR connected to the grid was that at Hunterston in 1976.

[3] AGR power stations comprise two reactors. Originally 14 reactors were built, 6 of these have been shut down and are being decommissioned.

© The Author(s) 2025
W. Nuttall et al., *Perspectives on Engineering Uncertainty*,
https://doi.org/10.1007/978-3-031-83254-3_4

Fig. 4.1 Dungeness B AGR Nuclear Power Station. *Image source* public domain via Wikimedia Commons (Sandpiper)

Whilst there are evolutionary differences in the detailed designs of the reactors, a general overview of the AGR design is given below. The differences in the details of the designs are not sufficiently large to affect the arguments and results presented elsewhere in this book. Further details of the design are given by Nonbøl (Nonbøl 1996).

A photograph of an AGR power station is given in Fig. 4.1.

A simplified, schematic, view is given in Fig. 4.2.

4.2 AGR Fuel

The fuel in an AGR reactor is in the form of uranium dioxide (UO_2) pellets with a diameter of 14.5 mm and containing a central annular hole. These pellets are encased in a stainless steel cladding tube 900 mm long. To compensate for the absorption of neutrons in the stainless steel the uranium in the pellets is enriched to 2.2–2.7% of the fissile U-235 isotope.

Thirty six fuel pins are fixed, by grids at the top and bottom and encased in two concentric graphite sleeves to form a fuel element. Eight fuel elements are then linked together with a tie bar to form a fuel stringer assembly. These fuel stringer

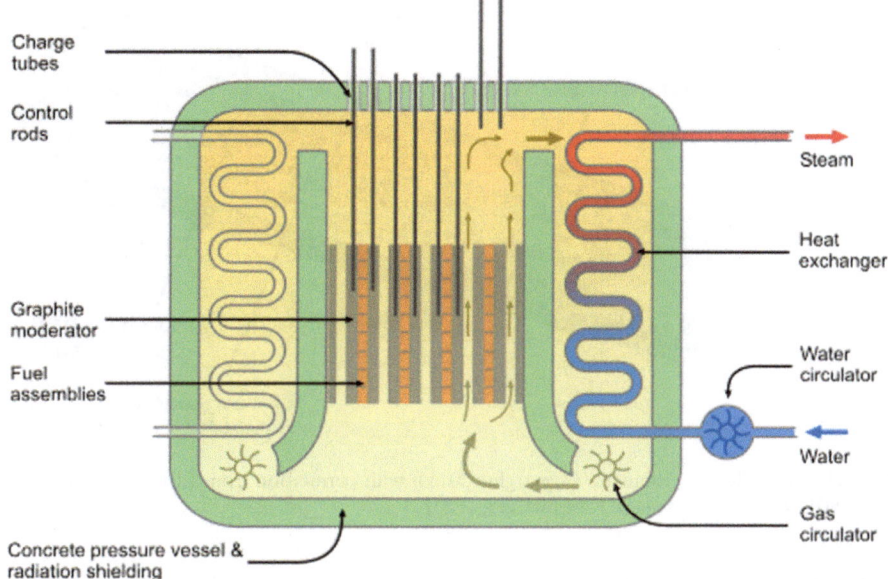

Charge
tubes

Control
rods

Steam

Heat
exchanger

Graphite
moderator

Water
circulator

Fuel
assemblies

Water

Gas
circulator

Concrete pressure vessel &
radiation shielding

Fig. 4.2 A schematic outline of an UK AGR. *Image source* Wikimedia Commons CC-SA-3.0
(MesserWoland)

assemblies are loaded into the reactor during initial fuelling or during refuelling
which takes place from the top of the reactor. The AGR was designed to be able to
be refuelled whilst at full power, however vibration problems mean that refuelling
takes place at reduced power, or when the reactor is shut down.

A fuel pin and fuel element are illustrated in Figs. 4.3 and 4.4.

Fig. 4.3 Part of an AGR
Fuel Pin. Copyright CC_
BY-4.0. *Image Source*
Thomas et al. (2018)

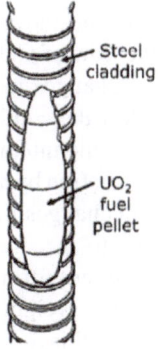

Steel
cladding

UO$_2$
fuel
pellet

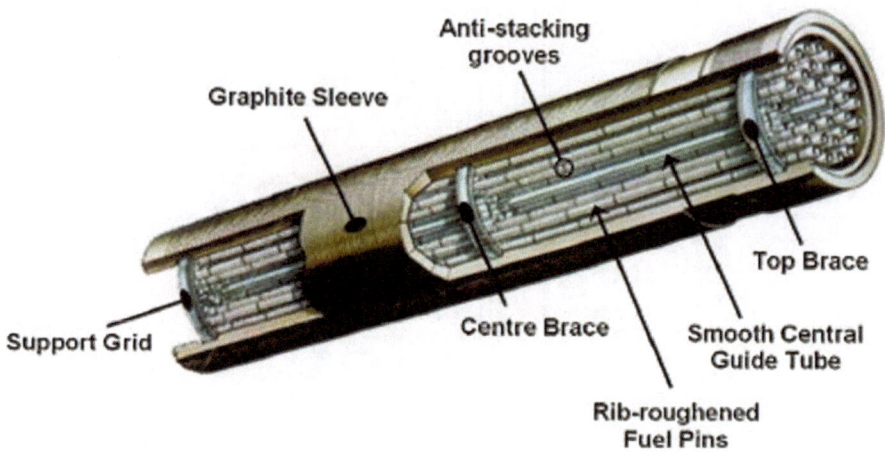

Fig. 4.4 An AGR Fuel Element. Copyright (2011), with permission from Elsevier Licence No. 5903780124108. *Image Source* Keshmiri et al. (2011)

4.3 The Reactor Core

The reactor core consists of an inner core of sixteen-sided graphite bricks. These are connected using graphite keys to give the core stability and to maintain the fuel channels in their correct spatial relationship. Within the core there are 308 to 408 fuel channels each containing a fuel stringer assembly. Control rod and instrumentation channels are also included within the graphite core. The graphite acts as moderator within the reactor as well as being a structural component. The graphite blocks containing the fuel stringer assemblies are surrounded by additional graphite blocks and by a steel shield. A section of an AGR reactor core is shown in Fig. 4.5.

Two of the AGR power stations i.e. four reactors (those at Heysham 2 in northwest England and at Torness in Scotland) also have outer graphite shielding blocks to direct the flow of CO_2 and shield components outside the core.

Changes in the properties of these graphite structures need to be understood to allow the continued safe operation of the reactor. The changes in the graphite result from the interaction of the core components with the CO_2 coolant and damage from irradiation by neutrons (ONR 2024b). Interaction with the coolant leads to weight loss and changes in the mechanical properties of the graphite. The dimensional changes due to the irradiation eventually lead to the formation of cracks in the graphite. Different forms of cracking can occur at different stages in the life of the reactor.

The weight loss, subsequent changes in the mechanical properties and cracking of graphite core components has the potential to affect the free movement of control rods and fuel elements and could also impact on the flow of coolant gas.

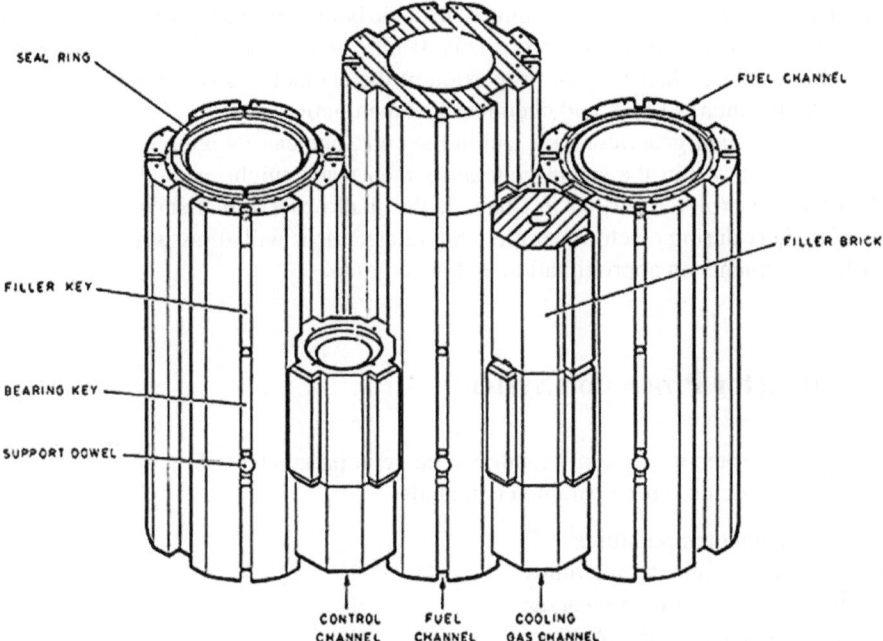

Fig. 4.5 AGR graphite core arrangement. *Image Source* (ONR 2022), https://www.onr.org.uk/media/au3hboio/heysham-torness-22-004.pdf. Contains public sector information licensed under the Open Government Licence v3.0

4.4   Reactivity Control

Control of the reactor is achieved through the use of control rods. These are made from stainless steel with 4.4% boron content, with boron a good absorber of neutrons. The reactivity of the reactor is controlled by raising or lowering of the control rods which are also used to shut down the reactor during normal operations and in an emergency.

4.5   Reactor Cooling and Heat Transfer

The reactor is cooled, and the heat transferred by the use of CO_2 gas. This gas is blown through the reactor using 8 electrically powered circulatory pumps located at the bottom of the core. As the gas passes up the reactor core heat is transferred from the fuel to the gas. A complete circulatory assembly with pump, impellor and guide vanes can be isolated from the gas circuit.

The CO_2 gas is fed into 4 steam generators located within the containment of the core. As the CO_2 gas flows through the steam generators it heats up water contained within boiler tubes. The water is pumped in at the bottom of the steam generators and exits as superheated steam at the top. This steam which is used to power turbogenerators and produce electricity. A proportion of the coolant is diverted into a by-pass loop which contains filters and chemical treatment units.

As these steam generators are within the core, one can imagine that if a failure were to occur within the steam generators then water might enter the core. Such ideas will be explored further in Chap. 5. We shall observe that such a scenario is most unlikely during reactor operation, but related ideas will allow us to make some further comments on approximation and uncertainty.

4.6 Reactor Protection System

The reactor protection system monitors a range of parameters of the reactor, and its associated systems. These parameters include:

- Fuel channel temperatures
- Coolant gas outlet temperature
- Neutron flux within the reactor
- Rate of change of neutron flux
- Electrical supply voltages, and
- Pump speeds.

If the system detects that any of these parameters have moved outside the permitted range it will initiate an automatic shut down of the reactor.

4.7 Shut Down Systems

If the performance of the reactor moves outside of the design parameters the reactor will be shut down, either by the operators or automatically by the reactor protection system. This is achieved by the rapid insertion of the control rods described above. If this fails to stop the chain reaction, nitrogen gas is introduced into the core, nitrogen being a better absorber of neutrons than CO_2. If this in turn fails, boron glass beads are released into the core. Boron is an effective absorber of neutrons.

4.7.1 Emergency Cooling

Following shut down of the reactor, cooling (to remove the decay heat) is provided by emergency cooling systems. When the reactor is still pressurised emergency coolant water is fed into one or more of the steam generators to remove the heat in place of the secondary circuit. If the gas circulatory pumps are not available natural, convective, flow is sufficient to provide cooling if emergency cooling water is provided to at least two steam generators.

If the reactor is unpressurised natural circulation is insufficient to provide adequate cooling and the coolant pumps must be used.

The AGR design represents a mature and well understood technology which has supplied electricity to the UK grid for nearly 50 years and past its original design life parameter of 25–20 years. As such, it gives a suitable example to illustrate the application of the principles described elsewhere in this book.

Acknowledgement This work was supported by the Engineering and Physical Sciences Research Council, UK via a grant entitled: *Enhanced Methodologies for Advanced Nuclear System Safety (eMEANSS)*. The grant had reference: EP/T016329/1. The authors thank the EPSRC for this support.

References

Keshmiri A (2011) Three-dimensional simulation of a simplified advanced gas-cooled reactor fuel element. Nucl Eng Des 241:4122–4135

Margulis M, Shwageraus E (2021) Advance gas-cooled reactor technology for enabling molten-salt design-estimation of coolant impact on neutronic performance. Nucl Eng Des 385

Mavrokefalidis D (2024) EDF plans to extend life of UK nuclear fleet 2024. https://www.energy livenews.com/2024/01/09/edf-plans-to-extend-life-of-uk-nuclear-fleet/. Accessed Date 29 Jan 2024

Nonbøl E (1996) Description of the advance gas cooled type of reactor (AGR)

Office for Nuclear Regulation (ONR) (2022) Agreement to NP/SC 7810 – Heysham 2 and Torness Power Stations, available at: https://www.onr.org.uk/media/au3hboio/heysham-torness-22-004. pdf

ONR (2024a) Operating Reactors. https://www.onr.org.uk/civil-nuclear-reactors/index.htm. Accessed Date 29 Jan 2024

ONR (2024b) Graphite core aging 2024. https://onr.org.uk/our-work/what-we-regulate/operat ional-power-stations/current-issues/graphite-core-ageing/#:~:text=During%20operation% 2C%20the%20graphite%20slowly,its%20effectiveness%20as%20a%20moderator. Accessed Date 3 Jul 2024

Thomas RN, Paluszny A, Hambley D, Hawthorne FM, Zimmerman RW (2018) Permeability of observed three-dimensional fracture networks in spent fuel pins. J Nucl Mater 510:613–622

Chapter 5
Uncertainty in Safe and Efficient Nuclear Power

5.1 Introduction

In Chap. 3 we introduced a set of important conceptual ideas around uncertainty. Some, while philosophically interesting, were judged to be of limited direct applicability. In that category we include quantum effects and deterministic chaos. We acknowledge quantum uncertainty in nuclear physics, but we do not see exploration of that aspect of physics as being fruitful for engineering safety and efficiency as discussed in these pages. The eMEANSS project had a subsidiary focus on nuclear data, from which uncertainties can be extremely important for reactor physics and nuclear engineering, but even in that respect we posit that the path to understanding does not lie via quantum mechanics per-se.

More profitably, we also discussed the importance of an uncertainty interval, and we expanded upon that to consider the importance of uncertainty in multiple dimensions (where we made an implicit assumption of orthogonality[1]). We also described how modern computational technology, and software enables complex uncertainty calculations and data visualisations.

In this chapter we step back and present a couple of illustrative case studies inspired by nuclear engineering realities and expressed with reference to our canonical nuclear reactor technology—the UK Advanced Gas Cooled Reactor, the AGR, as described in Chap. 4.

The eMEANSS project, introduced in Chap. 1, is devoted to safety considerations for civil nuclear power generation in two scenarios: normal operations and severe accident contingencies. In this chapter we shall present an illustrative idea drawn from each scenario space and we start with consideration of routine nuclear operations

---

[1] By 'orthogonality' we mean that the various dimensions of uncertainty are entirely independent of one another. For example, when seeking to understand the formation and growth of cracks in the graphite moderator of an AGR concerns around uncertainty in reactor temperature and neutron fluence might be regarded as separate and independent. Whether such parameters are indeed 'orthogonal' is something to consider carefully if uncertainties are not to be misestimated.

© The Author(s) 2025
W. Nuttall et al., *Perspectives on Engineering Uncertainty*,
https://doi.org/10.1007/978-3-031-83254-3_5

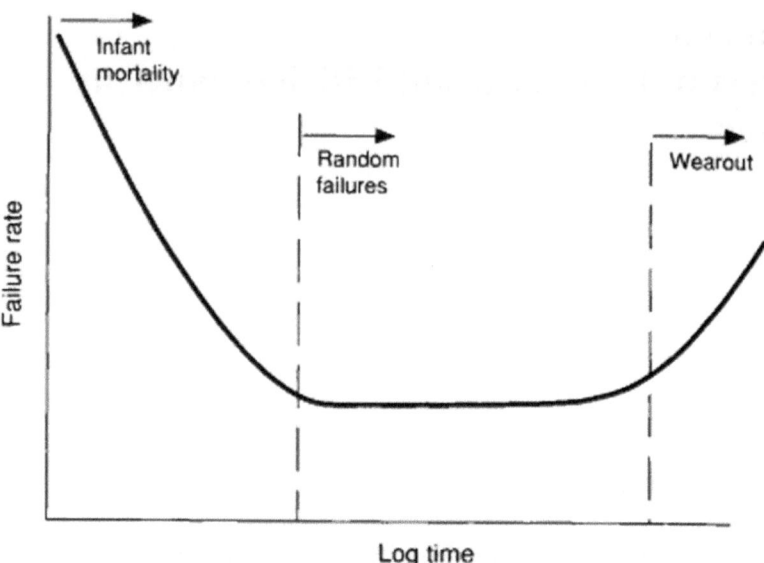

Fig. 5.1 Engineering reliability—the 'bathtub curve'. *Source* Ohring M. Engineering Materials Science: Elsevier; 1995. Reproduced under CCC Licence 5903720070280. All rights reserved

occurring over many years. Over such a long period of time safety problems can be growing and attention is needed to understand and mitigate the risks arising from emerging problems. As such we are, in our first case study below, considering the long-time effects exhibited by the 'bathtub curve' long invoked by engineering reliability scholars—see Fig. 5.1.

In essence the bathtub curve can be applied to any manufactured item required to function in a reliable engineering system. The representation shown in Fig. 5.1 corresponds to a scenario in which there is relatively little quality assurance and post-manufacture testing applied to the manufacturing process. In that scenario the deployment of the technology reveals many failures on initial use (termed 'infant mortality' in Fig. 5.1). To better understand this scenario let us imagine one of the simplest and most ubiquitous, but also most highly engineered items that we encounter in our daily lives. Let us focus on the humble ring-pull used to open a can of fizzy drink, or 'soda' as the Americans would say. Let us imagine these cans emerging from the manufacturing process and then being stored in a warehouse for 10 years. We start by considering the first few minutes. The scenario we are considering at this point relates to 'infant mortality'—one can expect that a proportion of the cans will not be sealed properly. It will be a small proportion, but it is not zero. It will be readily apparent when there is a problem. Indeed, any problem associated with the ring pull itself is likely to relate to a defect that causes the can to leak when first manufactured. These are the failures shown to the left of the bathtub diagram. Now we have all experienced the rare situation that a fizzy drink can fails to open—and perhaps the ring-pull comes off in our hands. That is rare and feels essentially random. It corresponds to the middle of the bathtub curve. Now let us imagine the cans after they have been in the warehouse for 10 years—and we are not here recommending that one should ever drink the contents of such a can. Our point is that after 10 years the ring pulls could be expected to fail more often than we see in our usual experience noting that the can has a polymer liner protecting the metal of the can from its acidic contents. This is an example of the degradation of reliability denoted by the label 'wear out' in Fig. 5.1.

Our first case example in the following section will fall into the scenario termed 'wear out'. One might say that given the stringent safety culture in construction that the 'infant mortality' phase is not often seen in the nuclear sector. We would suggest, however, that in principle the risk of 'infant mortality' in nuclear power technology is very real. Numerous nuclear power station projects have failed to complete construction owing to technical problems during the build. Frequently such technical problems combine with regulatory tightening and extended construction schedules undermining the economic case for the plant. Examples of civil nuclear power plants that suffered technical problems in construction and that died 'in infancy' are described in the text box below.

"Infant Mortality"

Examples of Nuclear Power Plants that Were Never Completed

One of the more recent examples is the failed project for units 2 and 3 at the VC Summer Nuclear Generating Station in South Carolina, USA. Construction of two Westinghouse AP-1000 reactors was abandoned by South Carolina Electric and Gas in July 2017 after reported expenditure of more than $ 9 billion (Collins 2018). The story involves cost escalation, missed deadlines and the bankruptcy of project partners. It is hard to point to any one single cause of failure except perhaps complexity and poor project management in all aspects.

Another US failed project worth mentioning is 1980s story of the Shoreham nuclear power plant built by Long Island Lighting Company (LILCO) of New York (Nuttall, 2022). The remarkable aspect of this case is that the power station was entirely complete when it was cancelled before commercial operations had started. The underlying problem in this case was again slow project progress. So slow that the plant was not yet in commercial operation when the Chernobyl nuclear accident in Ukraine occurred in April 1986. This accident raised concerns around mass population evacuation, something extremely difficult to imagine for an island location, like Shoreham NY. If the plant had managed to get beyond the zone of 'infant mortality' (Fig. 5.1) then it might have survived such challenges.

A similar story can be found in the Philippines—Bataan Nuclear Power Plant (BNPP) (WNA, 2024). As with Shoreham NY a completed nuclear power plant was never allowed to operate. In this case the Chernobyl nuclear accident also plays an important role in eroding public and government confidence in the project. Although the plant had been completed in 1984 it had not been commissioned by the time of the April 1986 Chernobyl accident.

The government ordered the project to stop prior to fuel being loaded (in contrast to the Shoreham experience). While there are real parallels between Shoreham NY and BNPP, in the BNPP case there were long-standing concerns around corruption and poor build quality during construction. Interestingly the site has been maintained for nearly 40 years giving rise to repeated speculation that at some point the project might be restarted. Realistically, however, one cannot posit that BNPP is a 40 year old infant—it seems more realistic to say that it died in infancy.

Finally, one cannot consider such failures to generate useful power without looking at the story of what is today known as Wunderland Kalkar (Wunderland-Kalkar, 2024). Developed on the site of a cancelled nuclear power station project, Wunderland Kalkar is a theme park and tourist attraction northwest of the German city of Essen and near the Dutch border; originally planned as the site of a completed, but never operated, sodium-cooled fast breeder reactor—the SNR-300. The SNR-300 was cancelled in early 1987, after construction was complete. The role of the Chernobyl accident in this story of failure is again clear. The extent to which the theme park has incorporated the buildings and structures of the original nuclear complex is interesting. The cooling tower, for example, is clearly visible at the heart of the attraction although now painted with jolly artwork.

Three of the stories above died in infancy in part at least to a random event far away and over which the project developers had no control—the Chernobyl disaster. This reality allows us to make an important point concerning infant mortality and the bathtub curve. It is not just that bad occurrences are more likely in infancy, the usual takeaway of the bathtub curve in safety engineering, it is that for nuclear power a random external adverse event occurrence can prove mortally wounding for those projects that are still in their infancy. As the bathtub curve suggests, early stage nuclear power projects are indeed much more likely to fail.

References

1. Collins J (2018) 1 year after nuclear plants abandoned, fallout continues. Associated Press. https://apnews.com/article/57a95fce520e4804941f585d7ec a97d6. Accessed Date 17 Sep 2024

2. Nuttall WJ (2022) Nuclear renaissance: technologies and policies for the future of nuclear power, 2nd edn. Routledge

3. World Nuclear Association (2024) Nuclear power in the Philip-ines. https://world-nuclear.org/information-library/country-profiles/countries-o-s/philippines. Accessed Date 17 Sep 2024

4. Wunderland Kalkar (2024) Marvel together. https://www.wunderlandkalkar. eu/en Accessed Date 17 Sep 2024

In the next section, we look to the other end of the bathtub curve, and we consider an issue of nuclear wear-out—the erosion in performance that might occur in the moderator of a British AGR power station. In the AGR design the function of neutron moderation is provided by graphite. Moderating graphite features prominently in at least two ways in the AGR design. Firstly, it is found within the fuel assembly (Fig. 4.4). A larger amount of graphite is found separately in specially engineered graphite blocks, or 'bricks', assembled in the core (Fig. 4.5). In this chapter we focus on Brick Graphite.

5.2 A Chronic Concern

For the AGR design, the propagation and accumulation of cracks in the graphite blocks used as the moderator could be a lifetime limiting process, as these blocks cannot be replaced. Notwithstanding the fact, however, that even a fully cracked core can be determined to be safe in certain circumstances. The rate at which cracking occurs is dependent on a number of factors each of which are uncertain. Consistent with the ideas introduced in Chap. 2, in this chapter we explain that such uncertainty can be better considered as an N-dimensional 'confidence space'.

The new approach describes how this probability space can be calculated and how derived results can be communicated to stakeholders. One might want to know what the number of cracks can be expected to be after, for example, 30 years of operations. Clearly, however good our modelling, theory and experimental testing, the number of cracks will be subject to uncertainty, as denoted by the error bars in Fig. 5.2.

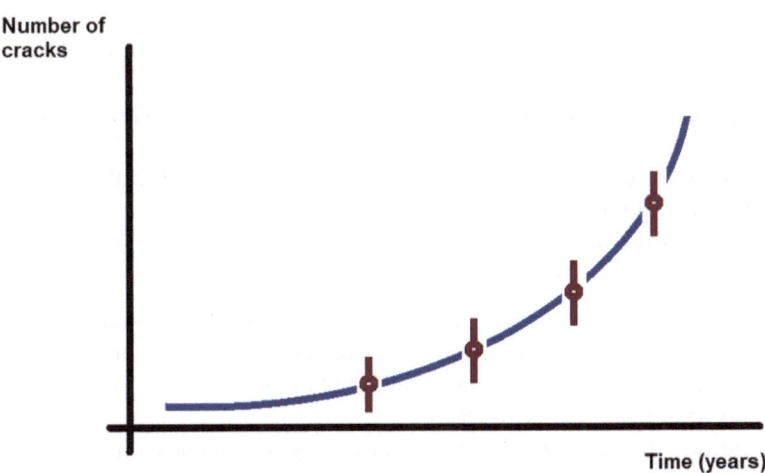

Fig. 5.2 Schematic representation of the number of cracks in AGR reactor graphite with time. *Source* used with kind permission of WJ Nuttall. All rights reserved

However, we must also think about the time axis in Fig. 5.2. In Fig. 5.2 time is being used as a proxy for hours of, full power equivalent, reactor operation. Whilst the reactor operator will have a prediction of how operating hours will increase with time, this prediction will be uncertain. Thinking more deeply, we can further note that it is not time that damages the graphite. Several considerations underpin the formation of graphite cracks. A key concern is the neutron irradiation, that is the underlying cause of the cracking of the blocks. It is not the rate of damage that matters (characterised by neutron flux) it is the accumulated damage that we are interested in (driven by neutron fluence). Neutron fluence is defined at the time integral of flux. It is measured in terms of neutrons crossing a planar surface (measured in m^2). There is no time rate involved in the definition, as there is with neutron flux. Fluence is a measure of accumulated irradiation, whereas flux is a measure of the radiation rate.

Neutron flux, and hence fluence, varies in a non-linear manner across a core, but it is modellable (at least theoretically) using modern reactor physics codes. The key factors in such modelling are the location of the control rods, the age of the fuel, and the power out of the core. Changes across the reactor (in space) contribute to uncertainty in the time evolution of flux (and hence fluence) at any given point within the core. The uncertainty of the neutron fluence links to uncertainties in the relationship between reactor power and neutron flux, etc., arising during the operating of the reactor.

Another key consideration that promotes the formation of graphite cracks in the AGR is weight loss from the graphite. The AGR design is based on carbon dioxide cooling and graphite moderation. At the high temperatures and pressures found in the AGR CO_2 cooling circuit, moderating carbon is removed to the gas phase by radiolytic oxidation and as a consequence the reactors exhibit mass loss from the moderator to the coolant.

Other relevant underlying uncertainty considerations concerning the number of cracks in AGR graphite include:

- Effects arising from the composition and manufacture of the graphite e.g. the porosity, grain size, and degree of graphitisation
- The operation of the reactor and the temperature and composition of the coolant gas
- The quality of codes used to model these interactions, but in this case the link is somewhat indirect and subtle. The quality of codes affects estimates through modelling of the number of cracks, and those estimates may affect operational choices that, in turn, affect subsequent crack production.

One can ask oneself to what extent these various factors represent dimensions of the problem (see Sect. 3.3). In this presentation, and at this stage, however, we shall keep it simple and stick with the issues of a 2-D Figure based on simply the number of cracks varying with time (Fig. 5.2).

The various effects at play in AGR moderator graphite during reactor operations have the consequence of initially putting critical regions of the blocks under compression, but then later the forces at these places on the blocks become tensile and are sufficient to cause cracking.

The neutron fluence will itself have numerous small sources of underlying uncertainty. Some of these will be epistemic uncertainties, but let us take the simplifying view that the dominant issue concerns irreducible aleatory uncertainties. These are based on other underlying aleatory uncertainties such as how well do we know the reactor power, or the position within the core? How does the power per unit volume and hence the neutron flux vary between different locations in the core? There may also be epistemic uncertainties that would also strengthen the critique, but the aleatory uncertainties are sufficient to remind us of the realities in play and that these realities are irreducible.

Such logic leads us to conclude that there is uncertainty in our understanding of brick graphite crack formation not only in the y-axis direction (number of cracks) but also in the x-axis direction (the relevant elapsed time, based on neutron fluence). This takes us to Fig.5.3, and by extension to Fig 5.4.

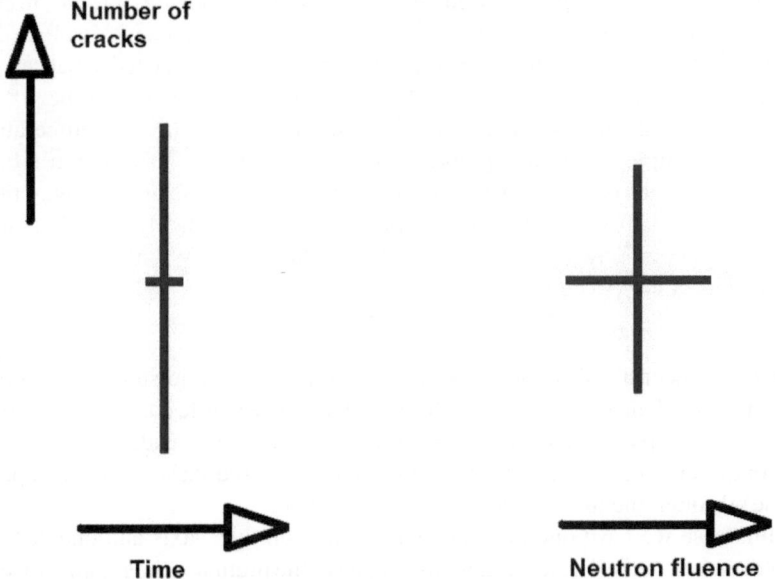

Fig. 5.3 Representation of uncertainties, not just in the number of cracks (y-axis of Fig. 5.2), but also in the x-axis variable as well—see main text. *Source* used with kind permission of the authors. All rights reserved

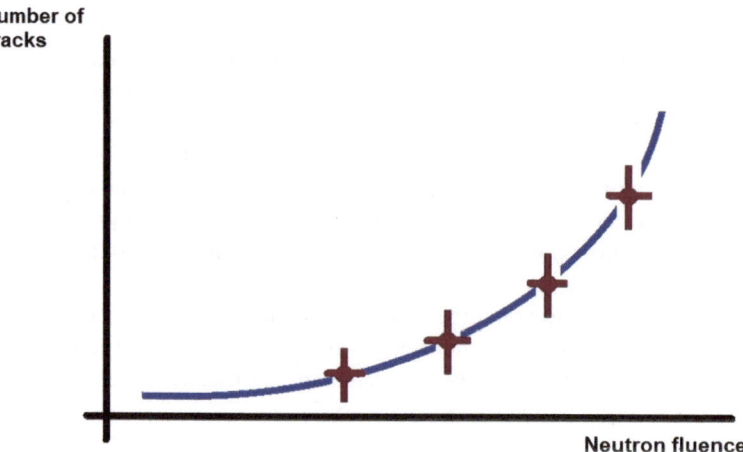

Fig. 5.4 Adjusted version of Fig. 5.2 showing two dimensions of uncertainty (error). *Source* used with kind permission of the authors. All rights reserved

---

Wigner Energy—An Aside

Finally, we must point out that an AGR power station operates its nuclear core at a high temperature (approximately 620 °C). As such the vibration of the carbon atoms in the graphite can anneal away microstructural damage caused directly by irradiation and to some extent AGR graphite can self-heal at the atomic level. Such materials physics is not sufficient to heal away the cracks referred to earlier, but it is sufficient to ensure that an AGR is free from the risks that caused the severe nuclear reactor accident at Windscale in Cumbria in October 1957. That was the result of the build-up of Wigner Energy in the low temperature core of a simple low power graphite moderated reactor. Operating at a high temperature any build-up of Wigner Energy (based on metastable interstitial carbon atom displacement) cannot happen as it is immediately annealed away at the high operating temperature found in an AGR core. Wigner energy is not a relevant safety concern in the UK AGR.

But each uncertainty is not simply a cross, it is better understood as an area in a two-dimensional uncertainty space. Here we need to be a little careful. In Chap. 3 we considered multiple dimensions of uncertainty. There we considered two, or more, orthogonal parameters and we represented the measured parameter (or dependent variable) through the use of colour, as in a heat map.

In this case we have one independent variable on the x-axis and one dependent variable on the y-axis. These variables do not have the mutual orthogonality discussed in Chap. 3, although for small deviations from the observed data point an orthogonality assumption can be made. It is clearly not the case that the actual underlying reality (as opposed to what has been measured in terms of the data point) must lie on the cross illustrated. It is clearly possible, indeed probable, that the true value is not on the cross at all. In essence it might be found anywhere in the blue shaded area of Fig. 5.5.

The error bars presented in Fig 5.5 express the idea of a confidence bound. For example, the y-axis error bars could represent the standard error of the mean values of the number of cracks at each elapsed time. They are not "absolute errors bars" guaranteeing that all the measurements lie inside the illustrated range. Looking at uncertainty in the x-axis direction, then if we think of time measurement, generally one might reasonably assume that such an error bar would be small and represent the precision of the measurement of the time-elapsed clock. That said, however, in this case the important physical issue is not time itself. As argued earlier, neutron fluence is the issue of greatest importance. The issue of graphite mass loss might, to a first approximation, be assumed to correlate closely with the fluence. Any error arising from such an assumption might be estimated and accommodated in the error bars deployed.[2]

The uncertainty space around the data point is assumed to be spanned by two parameters, in this case neutron fluence and number of cracks. We recognise that our understanding of fluence has an error associated with it and so does our understanding of the number of cracks. The actual neutron fluence and the actual number of cracks (as opposed to what we measure) is very likely to fall within the blue shaded area above.

The blue shaded area in Fig. 5.5 is deliberately drawn without a well described shape. We offer no hypothesis on which to propose a theoretically based proposal

Fig. 5.5 The actual uncertainty may be represented by a shape in a two-dimensional space denoting a region of confidence. *Source* used with kind permission of the authors. All rights reserved

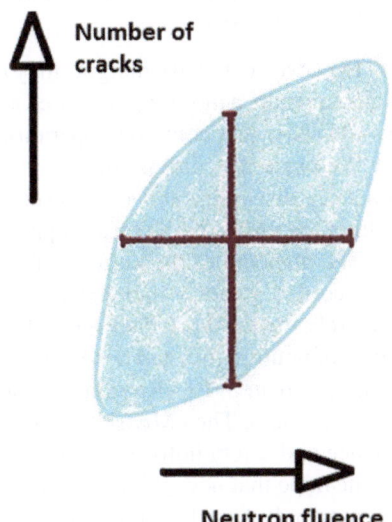

----

[2] Of course, one could continue to present the data (cracks, fluence, mass loss etc.) in a multidimensional plot scaled off elapsed time, measured to relative precision and accuracy, but the causal relationship between terms would not be obvious in such a data visualisation.

for the shape of that two dimensional representation. In a spirit of relative simplicity, however, it seems obvious to suggest that one can heuristically represent the uncertainty space in this case as a tilted ellipsoid. It is not known whether this suggestion is valid, it is merely an assumption that understanding of the number of cracks in nuclear reactor graphite as a function of time (or preferably neutron fluence) is better understood with such a probability space, than if one were to work with more naïve treatments such as error bars, or even bare data points (with interpolation).

In Fig. 5.4 we show a line connecting the various data points. This line-shape is as one might expect to be able to generate from a theory of crack formation and growth. Typically, such a theory would involve some unknown parameters that can be better understood via fits to the data. Fitting the theory to a data set characterised by uncertainty ellipsoids can be expected to yield a significantly more powerful insight than fitting to bare data points alone.

However, the issue for the reactor operator and the regulatory authority is not how much cracking will occur? The issue is "will the cracking be sufficient to impact on the structural integrity and will this prohibit operation of the reactor?" The relationship between the number and distribution of cracks and structural integrity introduces an additional uncertainty consideration. For the operator and regulator what has become an uncertainty volume in a theoretical space needs to be reduced to a very practical binary question "can the reactor continue to be operated or not?".

The UK's Office of Nuclear Regulation state in their Safety Assessment Principles (Office for Nuclear Regulation 2020).

> There should be an adequate margin between intended operational life and the safe working life of the graphite reactor cores. Safety margins should take due account of uncertainty in life predictions

However, extensive margins may mean that a reactor may be shut-down in circumstances where a more sophisticated approach to uncertainty analysis and combination would show that continued operation would be acceptable.

We close this discussion by observing that the two considerations (cracks and fluence) discussed above represented different Work Package technical domains within the eMEANSS research programme.

The question of the number of cracks in the graphite moderator of an Advanced Gas-cooled Reactor and the implications for nuclear operations and safety fall squarely within Work Package 2 (WP2) of the eMEANSS project. Meanwhile questions of neutron flux and hence neutron fluence (i.e. accumulated flux) links to issues considered in WP1 (Reactor Physics and Nuclear Data) and WP3 (Nuclear Fuel Performance). The eMEANSS project seeks to address issues nuclear power performance and safety holistically across all domains of nuclear science and engineering in the hope that new efficiencies may be found. In this stylised example we consider a case study based on the old AGR power plant concept. In reality eMEANSS is more future-oriented with WP2 considering TRISO-fuelled high temperature reactor concepts and WP1 and WP3 oriented to Light Water Reactor systems.

Having presented a discussion of the issue of graphite moderator cracks in UK AGR power stations it is important to make clear that the safety implications of such a scenario are not as serious as the reader may be assuming. Generally the moderator blocks key into one another. Even if cracked it is highly unlikely that the components of the block would move a significant distance even in an extreme scenario, such as an earthquake. While the loss of graphite mass can prompt concern that the reactor might be under moderated, the formation of cracks in the blocks is unlikely to be a direct cause for concern. There are two particular possible consequences of moderator block cracking that are worth consideration. One concerns operational performance and hence economics and the other is a safety issue. In both cases the concern is that a cracked graphite block shifts and blocks one of the designed channels giving access to the core. Similarly, debris from a cracked block might also block a core channel. If the cracked graphite blocks a fuel channel then, in-extremis, refuelling might be rendered impossible, hence ending the operational life of the reactor. A concern with direct safety implications is that a cracked moderator block might block the insertion of a control rod (grey rod) or shut-down rod (black rod). In all AGRs such a scenario would not prevent shut down of the reactor. As only a small proportion of shut down rods need to function for the reactor to shut down and furthermore there are always ample alternative options available (for example the injection of nitrogen gas into the coolant circuit). Looking beyond what has been presented in Chap. 3—the fundamentals of the AGR design are set out in (IAEA 2024).

Finally, again there is one other concern relating to cracks in the moderator of an AGR and that is coolant leakage within the core. Cracked graphite (and lost graphite mass) can open up gas flow opportunities within the core that do not follow the paths intended by the designers. This can erode cooling performance and hence operational performance of the reactor. It is unlikely to be a direct safety concern, it is much more likely to be an issue of inefficient heat transfer. For readers seeking to know more about the safety implications of cracks in AGR graphite then we recommend: (ONR 2024).

Finally we must stress that these various issues with graphite integrity in the UK AGR system have no consequences for the structural integrity of the power plant itself. The building and structure of the power station is not dependent upon the graphite structure within the reactor core.

One aspect of the eMEANSS project concerns the role of reactor operating temperature (e.g. WP1.4). This is a concern that can readily be appreciated with reference to the AGR design invoked for this book. The AGR is noteworthy for its high operating temperature—typically 625 °C in the core. To what extent are reactor neutronics and fuel and moderator structural integrity affected by such elevated temperatures?[3] If temperature effects are to be regarded as significant, then perhaps we should regard our graphical representation of the number of graphite cracks with effective time as being more multidimensional and not just 2-D, despite earlier comments. For example, consider the representation shown in Fig. 5.6.

---

[3] eMEANSS has separately considered nuclear data issues in nuclear engineering uncertainty—see the eMEANSS website (https://nubu.nu/emeanss/) for more information. Temperature can broaden spectral lines in nuclear data potentially altering reactor performance and/or operator understanding of such performance.

Fig. 5.6 The actual reality could be an N-dimensional space—denoted here for three dimensions. *Source* used with kind permission of the authors. All rights reserved

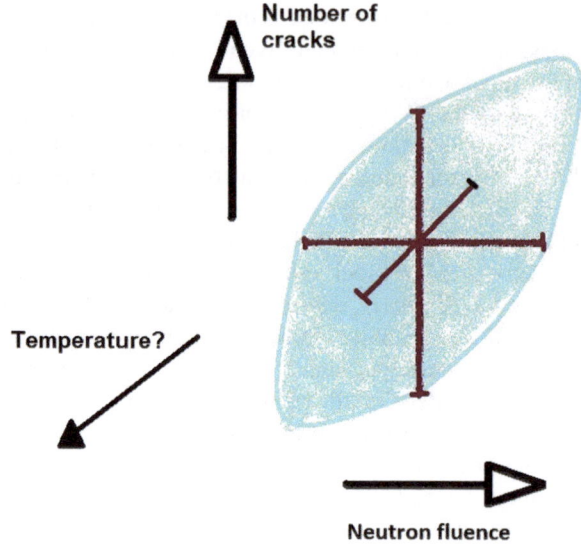

As an aside we should point out that while we earlier reported that the temperature of an AGR core is around 625 °C, that is a comment on the temperature of the coolant exiting the core and by extension an indication of the temperature of the graphite moderator, which it must be acknowledged varies in different locations in the core. The temperature at the annular centre of a nuclear fuel pellet would be substantially higher than that and typically far above 1000 °C.

While the issue of data visualisation helps us imagine multidimensional uncertainties, the reality is that the power of IT is relevant at another level. The processing of uncertainty problems involving epistemic and aleatory uncertainty in numerous underpinning and quasi orthogonal aspects is actually something that is computationally very intensive. The computer is not visualising as humans might, and it is as capable at managing a problem in five dimensions as it is in two. As modern affordable computing has grown in power, so has it become possible to imagine holistic uncertainty modelling and evaluation as proposed by the eMEANSS project.

The discussion above has focused on multidimensional uncertainties, for example in neutron flux and temperature. We have considered the impacts of such uncertainty with effective time and indeed noted that time is a proxy for neutron fluence and that there is also uncertainty in that parameter as well. Although we have considered the role of time, we have not yet considered issues of contingent probability.

Issues facing decision trees and the risks of cascading failure are very real in nuclear safety engineering. Nuclear safety case writers devote significant attention to the risks of multiple common cause failures and to the idea that failure can prompt failure and that a cascade to collapse must be avoided. eMEANSS is interested in such matters and the discipline of probabilistic risk assessment generally. The novelty of approach however is not so much in simply saying that linking simple safety challenges together can risk an emergent complexity at the system level. Rather we are looking for complexity at the individual component, or indeed conceptual, level and we seek to carry that uncertainty forward to holistic system understanding. As such, we are focussed on a collection of complex systems as a system—not that a system of simple components can show complexity.

Before leaving the topic of cracks in the graphite moderator of an AGR power station. Let us pause to consider how this example relates to some of the specific ideas and methods presented earlier in Chap. 3. In this chapter we have somewhat simplified a highly complicated reality. Modelling the formation and growth of a crack in nuclear graphite presents significant challenges due to the complexity of accurately capturing the crack's shape and behaviour (evolution over time). The exact shape and size of cracks are notoriously difficult to predict with precision, as they can be highly irregular and influenced by many factors within the nuclear reactor's environment, as discussed above. Such measurement uncertainties must be taken into account when attempting to model cracks. To address this challenge, one can use bounding intervals (see Sect. 3.5) to represent the possible range of crack dimensions, effectively capturing the imprecision of one's measurements. These intervals provide a conservative estimate of the crack's characteristics and help to ensure that any predictions about its growth encompass all possible scenarios. From this starting point, it becomes necessary to propagate these intervals through a mathematical or numerical model of crack formation and growth, ultimately producing a rigorous enclosure of the crack's potential evolution over time. If successful, this enclosure represents a guaranteed range within which the true behaviour of the crack must lie, provided that the initial input intervals accurately capture the uncertainty in the measurements. In the absence of precise predictive models, a Bayesian approach (see Sect. 3.7) may be used where the Bayesian prior is is updated by observations of cracking during the periodic inspections of the core (Maul et al. 2017).

Propagating intervals through a complex, computationally expensive non-linear model is a non-trivial task. The behaviour of cracks in materials such as nuclear graphite is often governed by intricate physical laws and complex interactions that can be difficult to model accurately. One approach to managing this complexity is sub-intervalization, a relatively simple method that involves splitting the original interval into n smaller subintervals (typically spaced linearly). The model is then evaluated independently for each subinterval, and the resulting ranges are combined (or "unioned") to produce an overall range that represents the possible evolution of the crack. This method relies on the assumption that within each small subinterval, the model exhibits a quasi-linear behaviour. While this assumption does not hold universally, there is always a sufficiently small subinterval for which the approximation becomes valid.

Sub-intervalization and similar interval propagation methods can help provide a robust estimate of crack formation and growth. This approach ensures that the computed range includes the true range of the uncertainty, as long as the initial intervals accurately capture the underlying uncertainty. The process is critical for ensuring that safety margins in nuclear reactors are maintained, and examples of such interval propagation techniques can be found where intervals are propagated through ordinary differential equations (ODE) modelling crack formation and growth, as outlined in reference (Gray et al. 2023).

In the example of cracks in AGR graphite blocks, we have considered system degradation with time using an example drawn from nuclear engineering. Nuclear engineers must consider similar aging effects across a range of technologies associated with nuclear power. Similar considerations apply, for example, to nuclear fuel performance and reliability. The most common form of reactor fuel in current use is uranium dioxide (UO_2). This is a refractory material with a high melting point. Using UO_2 fuels allows higher operating temperatures or increased safety margins at lower temperatures when compared to metallic fuels. When uranium atoms in the UO_2 are split two fission product atoms, along with neutrons, are produced. As these travel through the UO_2 they deposit energy through kinetic energy transfer to the surrounding fuel, raising the temperature. Understanding how this heat is transported through the UO_2, through the fuel cladding (usually zirconium or perhaps steel) and into the coolant is a challenge. Some fission products (the new elements produced as a result of the fission of uranium) are noble gases xenon and krypton. Due to their non-reactive nature, they tend to form bubbles, that act as a barrier to heat transfer and thus their build-up raises the temperature in the pellet (Qin et al. 2020). This increase in temperature increases mobility of fission products (Owen et al. 2023) and ultimately results in their release to the rod's free volume, raising its pressure and further reducing the heat transfer efficiency (Rest et al. 2019). In that paper one can see a feedback loop that ultimately prompts reactor designers to limit the approved fuel burn-up, or to restrict use of a fuel assembly after a given residency time or usage history in the reactor.

The transport of the fission products is directly linked to the micro-structure of the material, that is, the microscopic arrangement of the material's features. These features are often dictated by the manufacture process of the fuel and are therefore in-reactor properties which are affected by the manufacturing uncertainties. Conscious of these realities, various calculations of heat transfer have been performed based upon empirical assessments (Ronchi et al. 2004). They are affected by manufacturing uncertainties. For example, changes in the size of the fuel pellet by a small amount can have a large effect on the temperature at the centre of the pellet. Knowledge of this temperature is important as designers seek to improve efficiency while enhancing safety. Properties of the fuel which affect the transfer of heat change over time as the reactor is operated include, for instance, gas bubble formation in the UO_2. Understanding such processes is necessary if one is to understand how fuel performance changes over time.

5.3   Unpacking an Acute Scenario

In Sect. 5.2 we considered the uncertainty inherent in a safety critical metric slowly increasing with time. We used a medical analogy referring to the problem as being 'chronic'. However, nuclear safety must also consider the very fast moving events of a nuclear accident. If Sect. 5.2 is about the issues of a growing risk of component failure then this section concerns the systemic uncertainties associated with the dynamic processes of failure. As with Sect. 5.2, we explain the approach with reference to the AGR nuclear power station design.

In this section we start to return to the established notion of sequential effects. In a severe accident scenario the reactor is on a journey from early component failures to an incident involving the release of radioactivity to the environment. In such an extreme situation, the uncertainty in the amount of radioactivity released would reflect the uncertainties in the performance of the reactor fuel which is a major contributor to the source-term[4] uncertainty in a severe nuclear accident.

Uncertainty in the state of the reactor core and its fuel will reflect underlying uncertainties in, amongst others:

- The physics models used to describe the behaviour of the fuel elements. Including:

  - The models used to describe the formation of fission products and how they are affected by decay and by "burn up"
  - The parameters used in these reactor physics models.

  Beyond fuel modelling there are other concerns including:

- The parameters used to describe the composition of the material in the fuel rods
- The operational history of the reactor and how this has impacted on the fuel properties.
- How the reactor coolant behaves during the accident
- How other reactor components will behave during the accident scenario.

In seeking to find a topic suitable for us to make a pedagogical point concerning acute scenarios, we considered a set of design basis scenarios including a stuck control rod; a dropped fuel stringer; and a failure of the fuel's stainless-steel cladding. On balance, however, we judge that none of these scenarios is best suited to illustrating our main points of concern as we seek to provide the reader with an introduction to consideration of the role of uncertainties in nuclear engineering.

As we seek to make our methodological point, we soon concluded that we should attempt to unpack issues around a sudden insertion of reactivity. This type of fundamental consideration lies at the heart of very many acute reactor safety scenarios— whatever the reactor type. Reactivity insertion is, for example, a standard concern in the reactor physics of pressurised water reactor (PWR) systems. Within the general class of reactivity insertion, a canonical event of this type would be a rod ejection accident. While such an event could be taken to be a concern for the AGR design we feel there is a risk of causing confusion as the role and importance of such a scenario is rather different (and arguably less serious) in an AGR incident than it would be in a PWR. We kept looking and turned our attention to another scenario. As with all the scenarios we considered, it is an unlikely story and one that has been fully anticipated in the AGR reactor design. Again we found that we could not identify

---

[4] Also known as the 'accident release' the source term represents the isotopes of concern emitted from the site of a nuclear accident. In essence it is the total inventory of the release.

a plausible cause for concern relating to a reactivity insertion incident in the AGR design. In the next subsection we discuss a component failure sufficient for us to make a pedagogical point around rapidly moving events, but, as we shall explain, it is not a path to a severe accident.

5.3.1 Small Break Steam Generator Failure

The AGR steam generators are very high temperature CO_2 to water heat exchangers operating at high power density. Bongartz et al. report that for the now closed Hinkley Point B AGR there were two reactors and 24 steam generators accommodating a thermal power of 1500MW per reactor with an outlet gas temperature of 650°C (Bongartz et al. 1988). The steam generators are key pieces of advanced technology in the AGR design and they are highly complex structures involving, in particular, a very large number of high-performance welds. They are quite different from their equivalents in PWR systems. Let us imagine a weld failure potentially releasing steam into the reactor coolant circuit—this initiating event will lie at the heart of our acute scenario.

Bongartz et al. further report that the high (feedwater) pressure Hinkley Point B as having been about 160 bar and the second (reheat) pressure was about 41 bar. The coolant CO_2 pressure was also 41 bar. Hinkley Point B used serpentine boiler technology, rather than the pod boiler design used in earlier AGRs (Blackall 2019).

Blackall (2019) notes that fatigue and corrosion damage has required operators to plug individual tube platens (series of tubes through which steam flows) to prevent water escaping and coming into contact with the graphite core.

Nawal Prinja observes: "Boiler tube leaks are known to have occurred in the AGRs but there are no reports of boiler tube leaks leading to water ingress into the reactor core. The design of AGRs includes robust safety measures, such as physical separation between the primary carbon dioxide coolant circuit and the secondary water/steam circuit, as well as advanced monitoring systems to detect and address leaks promptly" (Prinja 2025). In addition, Bongartz et al. (1988) advise, based mostly on an observation of earlier Magnox reactor experience that for boiler tube failure generally: "A review of the data showed that the failure frequency is not connected with the load level (pressures, temperatures) or with the geometric size of the heating surface of the boiler. Design, construction, fabrication, examination and operation conditions have the greatest influence on the failure frequency, but they are practically not to be quantified. The typical leak develops from smallest size. By erosion effects of the entering water or steam it is enlarged to perhaps some mm, then usually it is detected by moisture monitors. Sudden tube breaks were not reported in the investigated period."

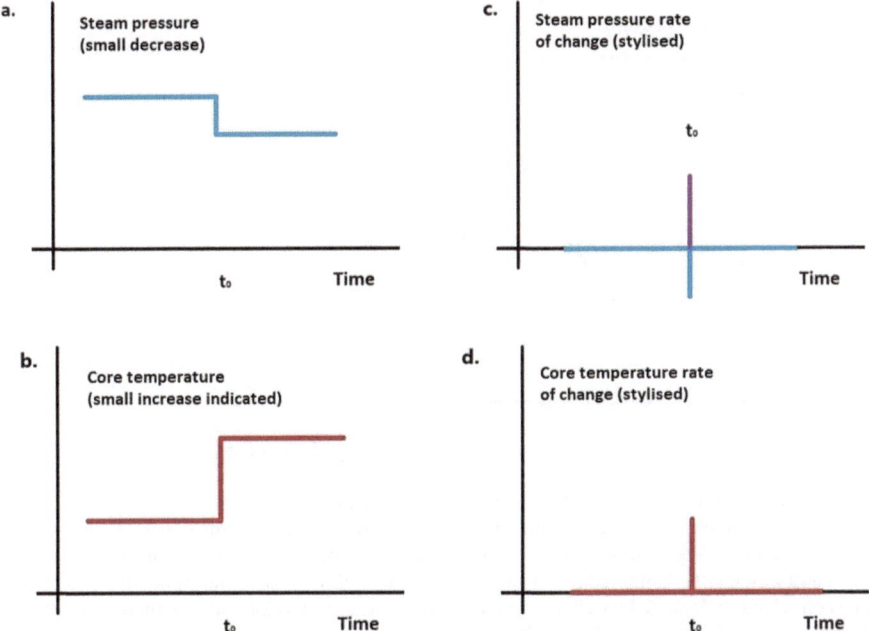

Fig. 5.7 Stylised representation of reactor state parameters and their time derivatives. In panels b. and d. we illustrate a temperature rise, but we do not posit the cause of the change – it is most unlikely to be reactivity insertion. *Source* used with kind permission of the authors.

We infer that the scenario of concern is a small leak across a low-pressure differential between the reheat steam and the CO_2 gas coolant. We assume a scenario of concern that does not risk a severe accident situation.[5]

For those seeking further insight into the design philosophy of the UK AGR and other gas cooled reactors we recommend (IAEA 1990). With the notion of a small water ingress event described for our reference technology, let us now imagine a similar situation in a future, highly instrumented, high temperature reactor with rapid data collection and storage capability.

Figure 5.7 illustrates such an event and its consequences in very simple terms with Heaviside jumps. Reality, however, is not best described using a Heaviside Unit Step Function (MIT 2024). The use of a Heaviside step is an approximation, in reality the rising edge will have structure and curvature. If we plot a rate of change a Heaviside step becomes a Delta function and once again that is a stylised approximation. In Fig. 5.7 panel c we simply map a negative Delta function to its positive mirror as that is probably how a reactor operator would see the data presented on their gauges.

---

[5] In an assessment prepared under the UK Radiation (Emergency Preparedness and Public Information) Regulations, EDF, as AGR power station operator, reports that generically for the AGR design 'boiler tube leak faults' is a potential factor precipitating a major release of radioactivity (EDF 2017). Less consequential risks are also discussed. Such formal assessment of remote possibilities does not undermine the logic presented above.

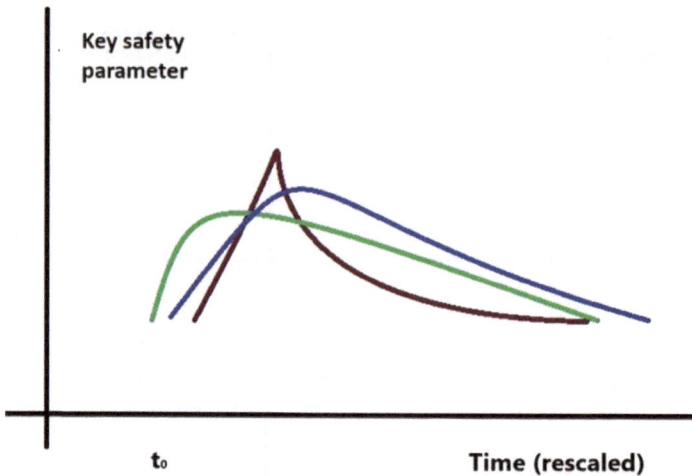

Fig. 5.8 In the future what might once have been considered to be a delta function might be monitored at high time resolution. Acute actions show various possible time dependencies—shown here schematically. The line shape has structure. One can expect that the line shape is composed of contributory elements combined in various ways: some additive, some multiplicative and perhaps some convolved. In Sect. 6.4 we shall consider some of these methodological elements. *Source* used with kind permission of the authors. All rights reserved

Figure 5.7 shows a stylised and simplified reality. With modern IT tools it now becomes practical to model the true line-shape of transient events without reference to such stylised representations. The naïve approach of modelling a transient derivative with a delta function (Fig. 5.7 panels: c and d) rather than a Lorentzian, or some other appropriate curve, can, miss important rate-of-change effects. For example Fig. 5.8 presents some illustrations of more realistic time dependencies that might be seen during the course of the event. For key safety parameters it can be instructive to look in detail at what happens during an acute scenario. These line-shapes have structure that can be established through the combination of theory and experiment and that can be easily represented using modern computational modelling tools. What tools and techniques might be brought in to understand such effects? We shall describe some relevant tools and techniques in the next chapter as we look ahead to future developments in the field of applied uncertainty research.

Acknowledgement This work was supported by the Engineering and Physical Sciences Research Council, UK via a grant entitled: *Enhanced Methodologies for Advanced Nuclear System Safety (eMEANSS)*. The grant had reference: EP/T016329/1. The authors thank the EPSRC for this support.

References

Blackall JL (2019) Modelling of in-line tube banks inside advanced gas-cooled reactor boilers, PhD Thesis, University of Manchester, UK (2019). https://research.manchester.ac.uk/en/studentTheses/modelling-of-in-line-tube-banks-inside-advanced-gas-cooled-reacto

Bongartz R, Breitbach G, Wolters J (1988) Frequency and distribution of leakages in steam generators of gas-cooled reactors, In: Technology of steam generators for gas-cooled reactors. IAEA International Working Group on Gas-Cooled Reactors (IWGGCR/15), pp 215–222

EdF (2017) 2017 REPPIR Report of Assessment: Hartlepool Power Station. https://www.edfenergy.com/sites/default/files/2017_har_roa_npm_-_final.pdf

Gray A, de Angelis M, Patelli E, Ferson S (2023) Bivarate dependency tracking in interval arithmetic. Mechan Syst Signal Proces 186. https://doi.org/10.1016/j.ymssp.2022.109771

IAEA (1990) Gas cooled reactor design and safety. Technical Report Series Vienna. Contract no. 312. https://www.iaea.org/publications/1416/gas-cooled-reactor-design-and-safety

IAEA (2024) General design and principles of the advanced gas-cooled reactor (AGR). https://nucleus-qa.iaea.org/sites/graphiteknowledgebase/wiki/Guide_to_Graphite/General%20Design%20and%20Principles%20of%20the%20Advanced%20Gas-Cooled%20Reactor%20(AGR).aspx. Accessed Date 16 Nov 2024

Maul P, Robinson P, Burrow J, Bond A (2017) Cracking in nuclear graphite. Mathematics Today, p 116. https://cdn.ima.org.uk/wp/wp-content/uploads/2017/06/Cracking-in-Nuclear-Graphite.pdf

MIT (2024) Step and box functions. https://ocw.mit.edu/courses/18-03sc-differential-equations-fall-2011/fda8bf925aa8c622c9c29c26aa346662_MIT18_03SCF11_s24_1text.pdf. Accessed Date 21 Nov 2024

Office for Nuclear Regulation (2020) Safety assessment principles for nuclear facilities, 2014 edn, Revision 1. Bootle 2020. https://www.onr.org.uk/publications/regulatory-guidance/regulatory-assessment-and-permissioning/safety-assessment-principles-saps/

ONR (2024) Graphite core aging 2024. https://onr.org.uk/our-work/what-we-regulate/operational-power-stations/current-issues/graphite-core-ageing/#:~:text=During%20operation%2C%20the%20graphite%20slowly,its%20effectiveness%20as%20a%20moderator. Accessed Date 3 July 2024

Owen MW, Cooper MWD, Rushton MJ, Claisse A, Lee WE, Middleburgh SC (2023). Diffusion in undoped and Cr-doped amorphous UO2. J Nucl Mater 576:154270. https://doi.org/10.1016/j.jnucmat.2023.154270

Prinja N (2025) private communication, text provided by email 11 April 2025 and approved for attributed publication 12 April 2025

Qin MJ, Middleburgh SC, Cooper MWD, Rushton MJD, Puide M, Kuo EY, Grimes RW, Lumpkin GR (2020) Thermal conductivity variation in uranium dioxide with gadolinia additions. J Nucl Mater 540:152258. https://doi.org/10.1016/j.jnucmat.2020.152258

Rest J, Cooper MWD, Spino J, Turnbull JA, Van Uffelen P, Walker CT (2019) Fission gas release from UO2 nuclear fuel: a review. J Nucl Mater 513:310–345. https://doi.org/10.1016/j.jnucmat.2018.08.019

Ronchi C, Sheindlin M, Staicu D, Kinoshita M (2004) Effect of burn-up on the thermal conductivity of uranium dioxide up to 100,000 MWdt^{-1}. J Nucl Mater 327(1):58–76. https://doi.org/10.1016/j.jnucmat.2004.01.018

Chapter 6
The Way Ahead

Although the standard approach, safety margins etc. (see Chap. 2) allows one to design safe systems and products it introduces several weaknesses.

The first and perhaps most important weakness is a tendency to 'overdesign'. A vaguely risk averse, or precautionary, attitude, if combined with a lack of quantitative understanding of the engineering system and its vulnerabilities, can lead to significant expenditure strengthening aspects of the existing design, but for little or no actual benefit. In this book we are keen to stress the importance of rigorous cost assessment linked the level of safety benefit achieved. Generally, instinctive tendencies to overdesign tend to produce designs that usually meet all requirements but which:

- are suboptimal in specific scenarios
- fail to account for all reasonable operational needs, and which
- frequently require extensive testing.

The call for extra testing arises when it becomes necessary, after the fact, to have a quantitative understanding of safety margins for a new configuration (see Sect. 2.1). Generally, the over-engineered design will have surpassed basic margins (often with much room to spare) but with changes planned those margins can no longer be trusted and proper evaluation is required. This further increases production costs, or in the case of the AGR increased cost of operations. We recommend early and rigorous analysis from the start so that all margins are clear, minimal for purpose and well understood. If it can be expected that future upgrading may be required to accommodate design changes then we recommend approaches found in the literature on Real Options and Engineering Flexibility, as discussed in Sect. 6.3.1.

Resilience—concerns those things that one cannot prevent because one does not know about the existence of the issue. One has no chance to quantify the frequency of such an event. One common heuristic is simply to insert a 'fudge factor' that means one might over-engineer the system by, say, 20%, to help its functionality ride out an event where it might be hit by something the project simply does not understand.

© The Author(s) 2025
W. Nuttall et al., *Perspectives on Engineering Uncertainty*,
https://doi.org/10.1007/978-3-031-83254-3_6

At some level this takes us back to consideration of safety margins, as discussed in Sect. 2.1. Furthermore, one might tolerate some level of failure, for example a brief degradation in performance that's temporary, but from which the system can easily and reliably recover.

Then we must consider the time domain, introducing notions of sequencing, causality and consequence. This takes us to established concerns around cascading failures. One principle very well established in nuclear safety is probabilistic risk assessment (PRA). PRA has long had an appreciation of cascading phenomena.

Time does matter—we would say that resilience comes with some limitations and some problems. How does one optimise for performance when such choices might compromise the safety of the system. Conversely if one optimises on safety is one's reactor still viable technically and economically?

Resilience thinking links to performance and safety. For example, even if the system exceeds tolerance in two out of five key parameters, it might still be safe. One must here think of correlated effects and anti-correlated effects. One must also ask whether there is a risk of common mode failures. Such issues are familiar in PRA.

A precautionary approach to safety in nuclear power can lead to a prompt shut down, but it must be recognised that a system in transition (to a shut down state, for example) is, in that moment, less stable and less inherently safe than when in continued normal operation. Change is itself can be more dangerous than persisting with normal operations.

A technology can be very successful with something that's not cutting edge, if it is used in the right way. If one optimises something that doesn't need to be optimised, then what value is really being added? In the next chapter (Sect. 7.2.2) we will consider examples from the railway industry including the British tilting Advanced Passenger Train (APT) of the early 1980s. In that case the design goal was to reduce journey time. To go from point A to point B railway industry strategists have, in essence, two competing approaches to consider. One can consider a new straight track and a more conventional train, or one can design a tilting train to cope with existing track curvature. In essence one can change the track[1] or change the train. A related consideration is the use of established infrastructure. At one level we take the existing infrastructure, as with the track in the APT case, but at another level is it merely sufficient to have a well-established infrastructure design that can be built reliably as in the alternative French TGV concept? We shall return to such thinking in Chap. 7.

In the case of nuclear energy, the twenty-first century sees a continued interest in light water reactors. One can ask why continue with light water reactors? They're not as thermodynamically efficient as later concepts, or even when compared to our case study example the UK AGR. The advantage of the PWR is that we know how

---

[1] Depending on the status of property rights in the country concerned, the rights of landowners may make changing railway tracks difficult or even effectively impossible. Arguably, the French (new track) approach was easier in France (for legal and constitutional reasons) than it would have been in the UK.

Table 6.1 ISO9000 Quality Management principles (WMP)

QMP 1	Customer focus	The primary focus of quality management is to meet customer requirements and to strive to exceed customer expectations
QMP 2	Leadership	Leaders at all levels establish unity of purpose and direction and create conditions in which people are engaged in achieving the organization's quality objectives
QMP 3	Engagement of people	Competent, empowered and engaged people at all levels throughout the organization are essential to enhance its capability to create and deliver value
QMP 4	Process approach	Consistent and predictable results are achieved more effectively and efficiently when activities are understood and managed as interrelated processes that function as a coherent system
QMP 5	Improvement	Successful organizations have an ongoing focus on improvement
QMP 6	Evidence-based decision making	Decisions based on the analysis and evaluation of data and information are more likely to produce desired results
QMP 7	Relationship management	For sustained success, an organization manages its relationships with interested parties, such as suppliers

The third column is the statement of each principle reproduced from source (ISO 2015)

to produce it. We know how to do it right. It's not just the fuel, we know how to do everything with hundreds of examples deployed worldwide. Such thinking quickly takes us to issues of 'quality' in engineering linking specifically to quality assurance and quality control. These concepts are now established in the form of standards, most notably the ISO9000 family of standards from the International Organization for Standarization. In particular, the ISO9000 framework established seven quality management principles (Table 6.1).

For the team designing a subsystem there is often a need to accommodate a particular uncertainty, in which case typically the team will find an engineering solution to do that. The team probably shouldn't modify everything or over-design everything. The team, however, is confronted with a parameter that is out of tolerance, and generally, as a consequence, the team needs to deal with that. Such realities can cause teams to act is ways that actually offer little to overall system resilience, or safety.

In summary we can say that all contributors to engineering system design need to characterize the uncertainty as seen at their level, based on their measurements and consideration of data that they have or know that they do not have. Then contributors (such as sub-system team leads) should pass that understanding of uncertainty all the way through their team-level modelling work. It is appropriate for the team to consider their own expert predictions and to reflect on confidence intervals. They should pass all that information forward and not only pass forward a single, deterministic values, even if with an error estimate.

It was suggested earlier (Sect. 2.2) that we need to have much better communication between individual component teams and the overall project leadership about all things, but especially about uncertainties. One reality to note is that in the past, for example in the 1960s or 1970s, such communication (at the level of functionals not functions) would effectively have been impossible. The computing technology (including all hardware, software, and networking) was simply not sufficient back then. Complex engineering systems exhibit rich uncertainty where vague, imprecise and probabilistic knowledge are present simultaneously.

In the nineteenth century and with the development of steam power, engineering systems emerged, but they lacked complexity and any instrumentation. It was possible for a steam engineer to describe the performance of the machine, up to and including its safety performance in a simple conversation with a colleague. In the years after the Second World War far more complex engineering systems emerged including nuclear power technologies. Typically, these devices were instrumented. As such the data exceeded that which could be explained in a free-standing conversation. Only now in the twenty-first century, have digital data management and communication improved to match the rich uncertainty of complex engineering systems. Artificial intelligence further adds to the prospect of enhanced digital communication.

In this book we should distinguish between the use of AI in nuclear system design and in nuclear reactor operations. The latter issue is the subject of some regulatory scrutiny and concern (World Nuclear News 2024), but it is the former that will represent our main concern.

6.1 Computer Systems and Uncertainty

Throughout this book we have referred to the enabling power of modern high-performance low-cost computing. Such tools permit the multi-dimensional probability spaces described earlier. They also enable the use of decision trees in risk estimation and a proper understanding of correlation and causation in safety engineering. In this way we are talking about the enabling power of modern IT in nuclear science and engineering, but there is another important issue to consider in the mid-2020s and that is the prospect for disruption caused by rapid innovation in IT. In particular the emergence in recent years of machine learning and generative artificial intelligence has the potential to change safety case writing, for example. Another concern lies in the area of nuclear security. The ability of large language model AI systems to draw upon very large amounts of scattered and sometimes obscure public information runs the risk that bad actors will, with the help of AI, find vulnerabilities in nuclear security. Clearly it is not in the public interest to engage in AI prompt engineering oriented to causing damage to a civil nuclear system, but we suggest that there is a threat here for policymakers and security specialists to consider. One can imagine AI software houses will be blocking certain key phrases before accepting AI prompts, but we posit that, initially at least, such protections will be of limited value and will be easily circumvented by well qualified and experienced experts seeking, for benign reasons, to boost their analytical efficiency via AI.

The relationship between safety and security is complex—in many ways the two concepts are mutually reinforcing. But one can imagine circumstances where they pull in different directions. A simple example would be the consequences of transparency. Generally, transparency is regarded as being aligned with safety, but equally transparency can prompt vulnerability and insecurity. Another example is the high radiation field from spent nuclear fuel. It is arguably a safety hazard while also being a route to enhanced security and non-proliferation as it makes diversion of spent fuel significantly more difficult.

The use of IT in nuclear system safety today increasingly involves the consideration of digital twins—virtual replicas of physical systems—that enable real-time monitoring, simulation, and predictive analysis using a combination of the many IT and uncertainty tools described in this book. Digital twins offer a powerful tool for enhancing safety by allowing operators to model and simulate the behaviour of nuclear systems under different conditions, identify potential risks, and optimize operational performance without direct intervention in the physical environment. This advanced technology aids in improving decision-making, anticipating system failures, and ensuring regulatory compliance, ultimately contributing to a more reliable, efficient, and safer nuclear energy infrastructure.

The CASL programme in the US developed digital models, twins as it were, that were capable of modelling PWRs. These models were validated against existing results, and allowed for the design, and consideration of new options, for reactor technologies moving forward. Examples of the codes are the BISON and Moose fuel performance code looking at how fuel evolves over time, both at the fuel pellet and fuel pin levels. With increasing computing power increasingly available at lower costs, this approach has the advantage of being able to deliver new information more rapidly than if a physical study is used. However, this approach needs to use validated modelling compared with experimental results. We shall return to digital twins in Sect. 6.2.

While risk and reliability assessment is as old as engineering itself, modern techniques emerged in the early 1960s from U.S. Cold War aerospace and missile programmes. Tasked with getting a man to the Moon and seeing the range of tools emerging, NASA did not focus on probabilistic risk assessment (PRA)—instead, the agency focused on Hazard Analysis (HA) and Failure Modes and Effects Analysis (FMEA). Separately, the adoption of PRA was led by the nuclear industry with particular progress being made in the 1970s (Fig. 6.1).

PRA is a systematic and quantitative approach used to evaluate and manage risks in complex systems. Unlike traditional risk assessment methods, which may focus solely on identifying potential hazards and their consequences, PRA goes further and incorporates the uncertainty of those hazards and consequences by applying probabilistic models.

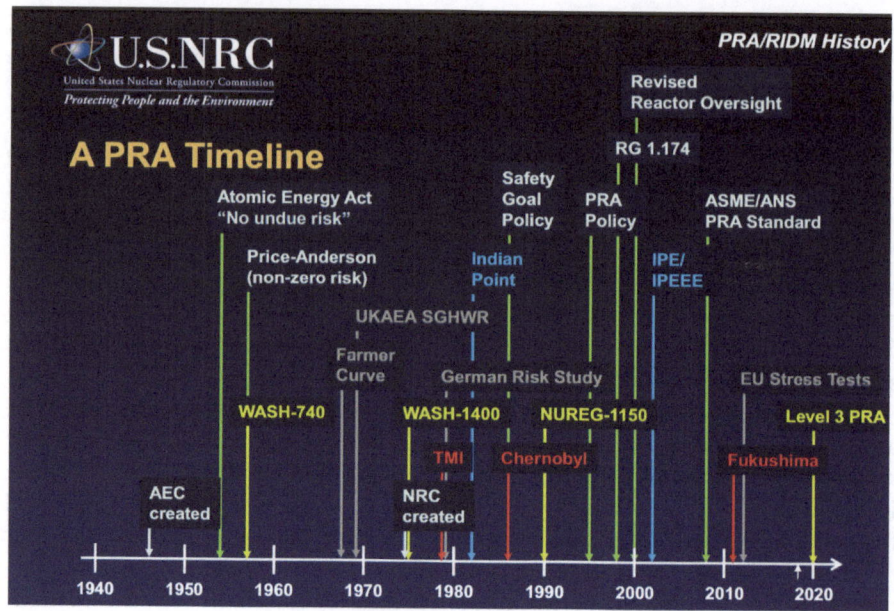

Fig. 6.1 Probabilistic risk assessment timeline. *Source* (US-NRC 2024) copyright US Government

At its core, PRA aims to estimate the likelihood of undesirable events (e.g., equipment failure, accidents, or environmental hazards) and their potential impact, taking into account both the frequency and severity of these events. This allows organizations to prioritize risks based on their probability and potential consequences, enabling more informed decision-making and resource allocation.

The process of PRA typically involves several key components:

1. **Hazard Identification**: Recognizing the potential hazards or failure modes that could lead to undesirable outcomes.
2. **System Modelling**: Constructing models (often fault trees or event trees) to represent the interactions and dependencies of system components.
3. **Quantification of Probabilities**: Estimating the likelihood of each event or failure mode using historical data, expert judgment, or statistical analysis.
4. **Consequence Assessment**: Evaluating the potential consequences of failures, including the impacts on safety, health, and the environment.
5. **Risk Evaluation**: Combining probability and consequence assessments to quantify the overall risk, often represented through metrics like risk matrices or risk curves.

By applying PRA, we can use the resultant analysis to better understand the range of risks they face, make more effective risk management decisions, and prioritise investment towards components where improved design can make the entire system more resilient to failures.

So far we have called for early and rigorous appreciation of risk and cost, but even within this approach there are weaknesses and gaps. For example, a reliable determination of failure probability can be extremely difficult, and erroneous assessments extremely consequential. Too often analysts retreat to the use of heuristic safety factors in an attempt to circumvent the need for proper assessments..

In the past decade, numerical simulations have set the foundation for the so-called analytical certification, using High-Fidelity models, though without adequately considering various uncertainties. An enhanced strategy is to take the best possible model and to add uncertainty on top of it. This is known as Best Estimate Plus Uncertainties (BEPU) (D'Auria and Mazzantinib 2011).

6.1.1 Probabilistic Design Analysis

Probabilistic Design Analysis (PDA) methods support the current deterministic engineering design practices to better understand design uncertainties to optimize safety factors and reduce conservatism (i.e. worst-on-worst design).

An enhanced strategy is to take various uncertainties well into account and, at the same time, replace High-Fidelity models with data-supported simpler models (surrogate models) that can be combined with Uncertainty Quantification (UQ). Therefore, the joint use of surrogate models and UQ methods offers a potential solution, as it addresses engineering problems in a cost-effective and technically viable manner. The crucial point is to ensure a comprehensive and accurate UQ. Additionally, the novel opportunities that Artificial Intelligence (AI) seems to offer in the domain pose specific challenges: (1) Ensuring accurate data for AI and Machine Learning (ML) techniques; (2) Estimating AI/ML technique prediction uncertainties; (3) Exploring AI/ML compliance with standards and regulations.

Surrogate Models are computational approximations used to represent complex and computationally expensive models. They serve as efficient substitutes, capturing the underlying relationships between input and output variables. By building a surrogate model, which is a simplified mathematical representation of the original model, one can significantly reduce computational costs while maintaining a reasonable level of accuracy. Surrogate Models are particularly useful for optimisation, sensitivity analysis, and uncertainty quantification tasks.

6.2 Digital Twins

Digital twins (DTs) are a fusion of digital technologies, in the sense that they can be considered as an evolution of simulators and models. In engineering, they refer to highly detailed virtual models that almost fully replicate physical objects, systems, or processes—connecting the virtual and physical worlds. These digital counterparts enable engineers to monitor, simulate, and optimize real-world assets within a virtual environment. By integrating real-time data from sensors and advanced analytics, digital twins can provide insights into performance, efficiency, and identify potential issues before they occur (predictive maintenance). This technology is already transforming industries like manufacturing, aerospace, and construction, allowing for the scheduling of maintenance before downtime occurs, improved design processes, and faster enhanced decision-making—ultimately all leading to reduced costs, increased operational efficiency and uptime, and improved safety.

One variant of a digital twin is to construct a software representation of an engineering design in order to study processes and optimise them before moving to physical real-world implementation (at far greater financial cost).

For a digital twin to be truly of use in decision making and simulation-based predictions, it must be representative of the uncertainties involved in the real physical processes, especially for large systems. Digital twins can be represented by a probabilistic graphical model which gives an integrated framework for calibration, data assimilation, uncertainty-informed decision-making, planning and control.

The digital twin acquires and assimilates observational data from the asset (e.g., data from sensors or manual inspections) and uses this information to continually update its internal models (e.g. Machine Learning models) so that they reflect the evolving physical system, which embodies a synergistic multi-way coupling between the physical system, the data collection, the computational models, and the decision-making process.

6.3   Thinking Broadly About Engineering Systems

In 2011 Olivier L. de Weck, Daniel Roos and Christopher L. Magee published a book outlining key concepts in technology scholarship and research that had emerged at the Massachusetts Institute of Technology and elsewhere over several decades. Their book entitled *Engineering Systems—Meeting Human Needs in a Complex Technological World,* explored a set of issues explained by the publisher (MIT Press) with these words:

> Today's large-scale, highly complex sociotechnical systems converge, interact, and depend on each other in ways engineers of old could barely have imagined. As scale, scope, and complexity increase, engineers consider technical and social issues together in a highly integrated way as they design flexible, adaptable, robust systems that can be easily modified and reconfigured to satisfy changing requirements and new technological opportunities. (MIT Press 2024)

In this book our focus is primarily on engineering system safety, but the lessons of de Weck et al. apply. The eMEANSS project observes how resilience and reliability, as assessed at a component level (for example nuclear reactor moderator performance), relates to the overall performance of a complex engineering system. The work of the MIT researchers reminds us that the engineering system is not just the nuclear power station as a functioning unit, but it goes wider to involve sociotechnical stakeholders such as electricity consumers, investors and national safety regulators.

Engineering systems as explored by de Weck et al. is not synonymous with Systems Engineering as understood by well-established bodies such as INCOSE (International Council on Systems Engineering). The systems engineering lens is more limited to the technical system as a complex challenge in engineering design management and operation. The scope of systems engineering would naturally be the entire nuclear power station, whereas, as we have outlined, the scope of engineering systems runs wider.

One consideration emerging from MIT's decades of work on engineering systems relates to understanding engineering uncertainties in terms of, not just uncertainty, but also flexibility. Building upon prior work by Richard de Neufville and Stefan Scholtes, Michel Cardin and coworkers set out the methodological basics of engineering systems research for a systems engineering audience in a paper subtitled: *A Methodology for Engineering Systems Design* (Cardin et al. 2007). More recently MIT has established a web-based resource to act as a knowledge hub for studies of this type, broadly termed 'strategic engineering'. This important resource can be found at: (Strategic Engineering 2024).

Running through this book there is a concern for the role of role of emergent information in understanding system safety. Fundamentally there are two ways in which to assess safety in systems engineering. The first is the focus that we have tended to adopt in this book, namely that one builds up understanding from an understanding of subsidiary units and issues. In this book we have discussed how this might be done in more sophisticated ways assisted by modern IT systems. The other approach, however, is philosophically quite different. While the former idea seeks to theorise and construct an understanding of system safety the second approach is grounded in observation—how, and why, do systems fail?

The latter approach has been enormously successful in improving the safety of commercial aviation, but it must be recognised that nearly 30,000 commercial aircraft are in service around the world. Sad to say, and as noted earlier in Chap. 1, such a large fleet gives a sufficient number of accidents and near-misses for safety progress to be made in realistic timescales. Nuclear power is rather different. The total number of in-service nuclear power reactors is roughly 440, noting that some power stations will have more than one reactor on-site. Nevertheless, the number of reactors is only approximately 1% the number of aircraft in commercial service. Arguably, this implies that if improvements in nuclear safety were solely as a result of "lessons learned" that safety would only improve at 1% of the rate that aviation safety improves. That said, however, nuclear power is under severe public policy pressure worldwide to improve system safety. That level of effort is arguably more than the economic base of the, still relatively small, industry can support.

Given the challenges inherent in taking safety improvement from the observation of failures, scholars of nuclear power plant performance sometimes invoke reliability as a proxy for safety. A more general pointer to such a logic is discussed here: (Moore 2024). The concepts of reliability and safety are quite distinct, but the point is that arguably the two correlate with one another—a *reliable plant is a safe plant* etc.

If one accepts such logic then adopting an MIT-inspired engineering systems mindset allows one to make a link from the world of eMEANSS to the arena of nuclear power finance and project economics. Such a leap to metrics measured in terms of costs and revenues opens up the possibility of linking to relatively sophisticated insights from finance. Such issues will be explored briefly in the next subsection.

6.3.1   Real Options and Engineering Flexibility

The concept of Engineering Flexibility describes the practice of designing and constructing a plant with the necessary facilities to be retrofitted so that alternative products can be supplied to the markets. Including the capability to be retrofitted at the construction stage incurs additional costs for the developer. This additional cost may be worthwhile to the developer as it provides the ability to respond proactively to changes in market or regulatory environments. Trigeorgis, has stated that such Engineering Flexibilities result in Real Options which give "the potential to conceptualise and quantify the value of options from management and strategic interactions" (Trigeorgis 1996). These Real Options can be regarded as being analogous to financial options and have an intrinsic value (Neufville and Scholtes 2011).

Cardin (2013) has described a strategy for identifying and incorporating flexibility into an engineering design. This strategy consists of a number of stages:

- Initial baseline design of the plant
- Recognition of uncertainties that may affect the economic performance of the design
- Generation of design concepts that may allow adaption to respond as the identified uncertainties are resolved
- Use of quantitative analyses to identify which Engineering Flexibility would give the best value.

Amram and Kulatilkala (1999) have identified a range of methods for valuing Real Options. These include the use of binominal lattices (Copeland and Antikarov 2003) and Monte-Carlo modelling techniques.

Real Option studies have been carried out related to nuclear energy. Shi and Song (2013) have shown that the use of Engineering Flexibilities would increase the value of a nuclear power programme in China. Locatelli and colleagues (Locatelli et al. 2014) have shown that a "wait and see" flexibility adds value to a small nuclear reactor project (SMR). Separately, Locatelli et al. (2017) have shown that the Engineering Flexibility to switch between electricity generation and desalination at times of low electricity prices adds value to a SMR. Similarly, Cardin and colleagues have shown that a Real Options approach to a nuclear power programme can give additional value through the use of the flexibilities of staging, expansion and life extension (Cardin et al. 2017).

Linkages between engineering flexibility, management under uncertainty and the challenges and opportunities of nuclear energy have been explored in an earlier Indo-UK Civil Nuclear Research Partnership collaboration known as 'NREFS'. The project *Management of Nuclear Risk Issues: Environmental, Financial and Safety* yielded a series of research outputs which can be accessed at no charge via the project website (NREFS 2024).

6.4 Advanced Methods

In this section we provide pointers to some more advanced uncertainty methods now being applied to problems in nuclear engineering.

6.4.1 Petri Nets and System Dynamics

Petri nets, a product of Carl Adam Petri's doctoral thesis (Petri 1962) are a graphical network system used to represent visually a system in which there are multiple independent processes active in parallel at the same time.

Petri nets consist of two kinds of nodes: *places* and *transitions*. *Places* represent conditions within the system under modelling, while *transitions* represent events that may cause the condition of the system to change. *Input arcs* connect places to transitions while *output arcs* connect transitions to places, with both being one-way connections. Small graphical markings called *tokens* are drawn within places to visualize the 'flow' through the network. A transition may 'fire' if there are enough tokens in each of the transition's input places. Upon firing, the token is consumed, and tokens are placed within the output places.

By modelling systems with Petri Nets, we can better understand the interaction of concurrent subsystems and processes, something which is commonplace in process engineering systems such as nuclear power plants and, in the context of operational safety, we can better visualise the flow of emergency procedures under non-nominal situations.

Intuitive Example of a Basic Petri Net

Consider a simple system which models a kitchen chef. The chef has two states, they are either waiting for an order, or they are preparing an order—these are the two possible states of our system. To change the state of the system, an order must arrive, or the order must have been completed—these are the two transitions of our system. The resultant petri net is as in Fig. 6.2.

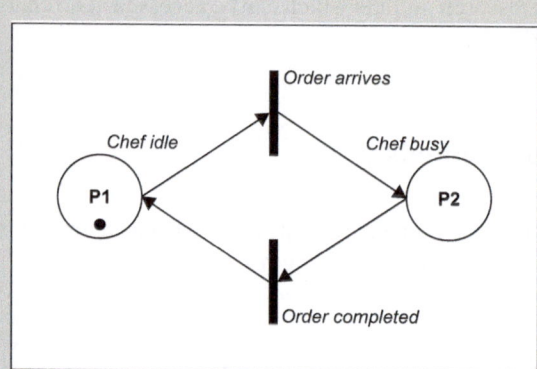

Fig. 6.2 Simple Petri Net—Chef's work scheme in the kitchen. *Source* used with kind permission of Ewan Smith. All rights reserved

In Fig. 6.3 we provide a related analysis applied to a nuclear power plant. Although this diagram was not originally developed with the Advanced Gas Cooled reactor in mind, most of the relevant considerations would apply to the AGR design introduced in Chap. 4. Similarly in Fig. 6.4 we provide a Petri Net developed for Light Water Reactor systems. Figure 6.4 is presented in terms of System Dynamics a powerful tool developed by Professor Jay Forrester and colleagues at the Massachusetts Institute of Technology in the 1950s. System Dynamics is now a well-developed approach to complex problems with a strong research community (Systems Dynamics Society 2024).

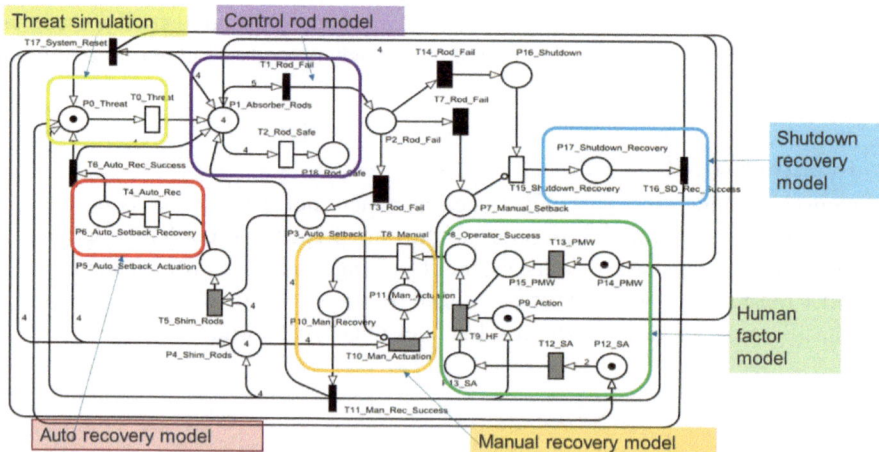

Fig. 6.3 Petri Net Model of a nuclear power plant. *Source and copyright* T. V. Santhosh and Edoardo Patelli, with kind permission, all rights reserved (Santhosh and Patelli 2021)

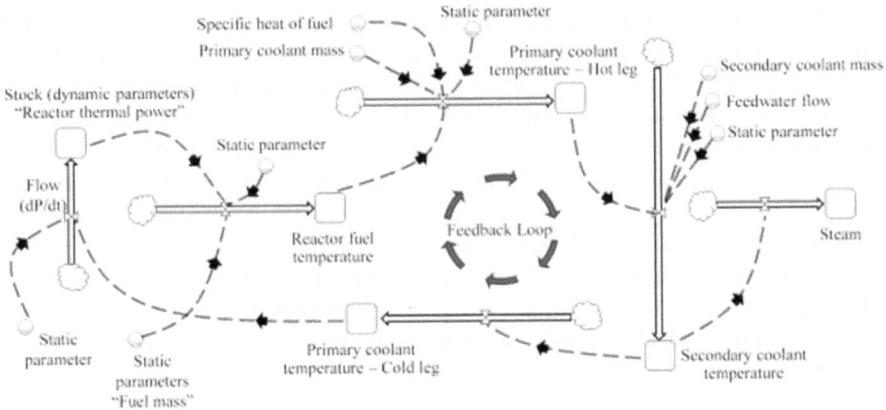

Specific heat of fuel

Static parameter

Primary coolant mass

Primary coolant temperature – Hot leg

Secondary coolant mass

Feedwater flow

Stock (dynamic parameters) "Reactor thermal power"

Static parameter

Static parameter

Flow (dP/dt)

Reactor fuel temperature

Feedback Loop

Steam

Static parameter

Static parameters "Fuel mass"

Primary coolant temperature – Cold leg

Secondary coolant temperature

Fig. 6.4 Petri Net model of PWR nuclear power plant as implemented in Machiatto Software from the University of Nottingham. *Source* https://doi.org/10.1016/j.net.2019.04.017 CREATIVE COMMONS

6.4.2 Automatic Troubleshooting Processes

Automatic troubleshooting processes in engineering involve the use of technology, such as artificial intelligence (AI), machine learning, and automation, to diagnose, analyze, and resolve technical issues without human intervention. These systems are designed to identify problems quickly, assess potential causes, and implement corrective actions based on predefined rules or learned patterns. The core goal is that by automating the troubleshooting process, engineers can minimize downtime, reduce human error, and enhance efficiency in system maintenance and repair by cutting out the need for human response. This approach is particularly valuable in complex systems, where the ability to detect and resolve issues in real-time is critical for maintaining operational continuity and safety. The recent advent of text generating AI means a domain-trained bot can generate post-incident reports for human operators to understand and to allow them to research process improvements which can avoid future incidents reoccurring. At the time of writing the role of AI in nuclear engineering and safety management is very fast moving. Rather than point to the fast moving literature, we recommend the insights of Dr Nawal Prinja author of the Foreword to this book. We shall also return to AI and its importance in Sect. 6.4.5.

6.4.3 Bayesian Approaches and Credal Networks

Bayesian networks (also known as Bayes nets, Belief networks, and Causal networks) are a practical, intuitive approach to building probabilistic models, utilising graphical nodes and the connections between them, i.e. a Directed Acyclic Graph (a graph with directed links and no directed cycles) and they can be used to solve any causal models (e.g, Fault tree see (Medkour et al. 2017)).

Each node in a Bayesian network represents a variable within the model, which may be discrete (a set of mutually exclusive states such as City = [Glasgow, Bangor, Oxford, …]) or continuous (a probability distribution). Each node is connected to other nodes via one-directional links which allow for one node to influence the other. While it may be that not all nodes are directly connected to each other, they may still be dependent upon each other as they may be connected indirectly through other nodes.

With a fully setup network, inferences can be drawn from the network on the result of interest. For instance, following the example in Fig. 6.5, if we are interested in getting to work on time, then, in detail, we are interested in the joint probability of being on time, our alarm being on, that we do not oversleep and that our bus is not late:

$$P(\text{On time, Alarm on, No oversleep, Bus on time})$$
$$= P(\text{Alarm on}) \cdot P(\text{Not oversleep} \mid \text{Alarm on}) \cdot P(\text{Bus not late})$$
$$\cdot P(\text{On time} \mid \text{Bus not late, not overslept})$$
$$= 0.9 \cdot 0.9 \cdot 0.8 \cdot 0.9 = 0.5832 = 58.32\%$$

Credal networks are an extension of Bayesian networks designed to handle epistemic uncertainty within the probabilistic model robustly. Unlike Bayesian networks, which rely on precise discrete probability values/continuous distributions, credal

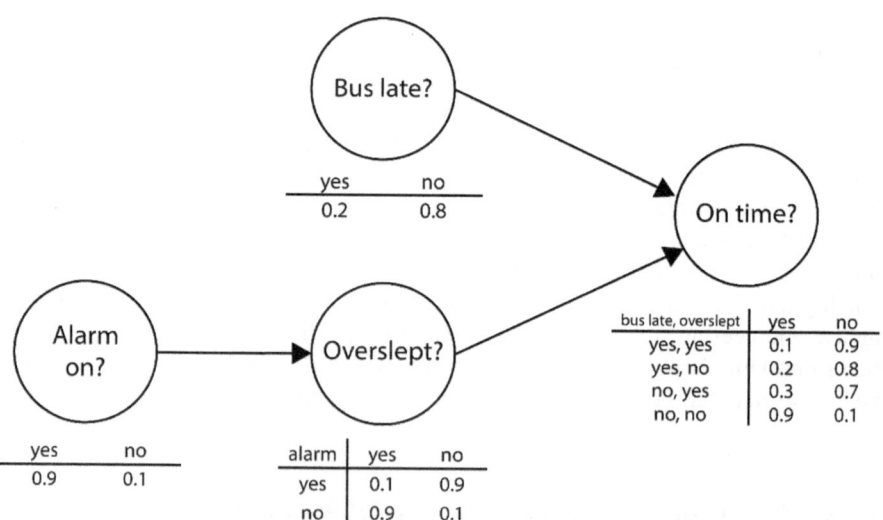

Fig. 6.5 A simple example of a Bayesian network, used to determine the probability of arriving to work on time. The probability of the alarm being turned on influences the probability of oversleeping, while the probability of oversleeping and the probability of the bus being late both influence the overall probability of arriving on time. *Source* used with kind permission of Ewan Smith– all rights reserved

networks allow for imprecise or interval-valued probabilities, making them ideal for situations where information is incomplete or uncertain, enabling epistemic uncertainty to be considered within the analyses. This flexibility is achieved by representing probabilities using sets of distributions, called credal sets, instead of fixed probability mass functions. By doing so, credal networks provide a more conservative approach to decision-making, as they account for a wider range of possible outcomes, representing a deeper level of epistemic uncertainty in the model which better reflects the current knowledge of the system, adding confidence to decisions made with such networks (Fig. 6.6).

Alternatively, we can look at Credal networks from a slightly different view, consider a football match where 4 bookmakers are offering odds on two teams, Glasgow United FC and Bangor City FC, winning or not winning (notation: $\neg$) their respective matches this weekend (Table 6.2).

Taking the bookmakers odds as what they consider the probability of each outcome to be, you could form a Bayesian network by bounding the probabilities with intervals, for example, $P_i(G, B) = [0.25, 0.75]$.

For further insight into Credal Networks as applied to nuclear engineering we recommend the work of Estrada-Lugo and colleagues (Estrada-Lugo et al. 2020) and references therein.

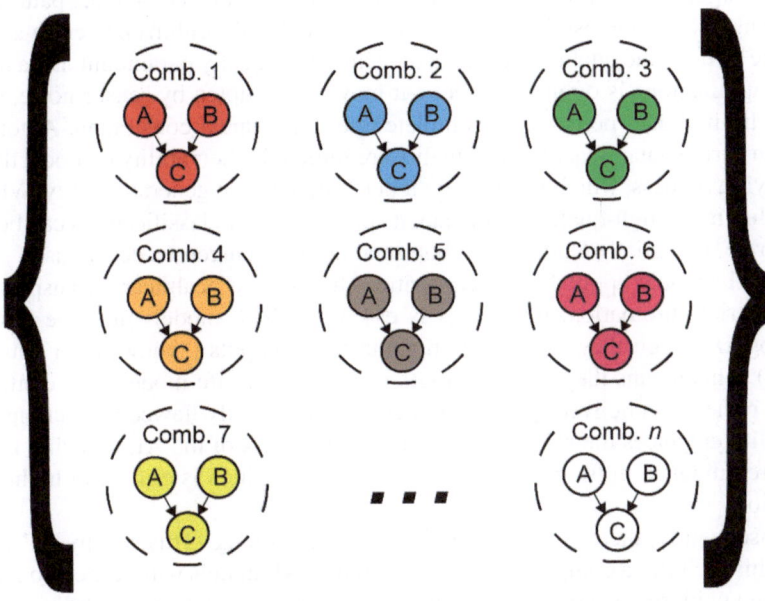

Fig. 6.6 A simplified view of a Credal Network's architecture. The credal set is a finite set of Bayesian networks, each with its own unique and specific combination of probabilities. *Source* used with kind permission of Ewan Smith. All rights reserved

Table 6.2 Example of a credal network, considering probabilities of a team's win from across 4 different bookmakers. See the main text for notes on notation

Bookmaker	$P_i(G, B)$	$P_i(\neg G, B)$	$P_i(G, \neg B)$	$P_i(\neg G, \neg B)$
bet247	12/16 (0.75)	1/16 (0.0625)	2/16 (0.125)	1/16 (0.0625)
Ladybrooks	12/16 (0.75)	2/16 (0.125)	1/16 (0.0625)	1/16 (0.0625)
Betunfair	4/16 (0.25)	1/16 (0.0625)	7/16 (0.4375)	4/16 (0.25)
Betfrank	4/16 (0.25)	7/16 (0.4375)	1/16 (0.0625)	4/16 (0.25)

6.4.4   Challenges in Deploying AI-Based Advanced Methods

Artificial-intelligence-driven decisions and the underlying algorithms must be explainable to foster trust and demonstrate resilience against real-world challenges. They must handle noisy data from various measurement devices and continue functioning effectively, even if the data streams are compromised, tampered with, or interrupted. Ensuring that these algorithms are both transparent and adaptable builds confidence among stakeholders, who must rely on the AI's accuracy and stability under fluctuating conditions.

For example, an implementation of digital twin technology may leverage machine learning algorithms to process vast quantities of sensor data, recognizing patterns and providing actionable insights in real time.[2] While the algorithms powering a digital twin are critical, the quality of the data they ingest is equally paramount. Data in real-world applications is often imperfect—it may be corrupted by sensor noise, sparse due to limited data points, or incomplete due to technical constraints. Algorithms that learn from data of insufficient quality are limited in their ability to model the true underlying process, which hampers their performance and generalizability. When an algorithm lacks high-quality input data, its predictions or classifications can become inaccurate, potentially leading to suboptimal or even dangerous decisions.

For AI models applied in safety–critical fields, like healthcare, transportation, or industrial automation, robustness is essential. Such models must be designed to recognize when they are uncertain about their outputs ("know when you don't know"). This means they can alert users or revert to a safe mode when confidence is low, rather than delivering overconfident but potentially flawed predictions. This capability enhances the robustness and trustworthiness of the AI, reducing the risk of overconfidence in the model's decision-making, which is critical in high-stakes scenarios.

Consequently, there is a growing focus on developing 'Trustworthy AI' principles. This includes a comprehensive evaluation of AI models with respect to fairness (ensuring unbiased outcomes across diverse demographics), interpretability (making model decisions understandable to users), and robustness (ensuring stable performance under various conditions). Trustworthy AI is foundational to building AI systems that are not only technically proficient but also socially responsible, fostering broader acceptance and integration into society. The reliability of these systems must be actively implemented by means of scientific methods that take explicit account of the uncertainties. This implementation faces significant challenges, including managing uncertainty within the data, verifying the accuracy of scientific numerical

---

[2] See e.g. discussion in Sect. 5.3.1 around Fig. 5.8.

calculations, validating simulation models against empirical data, and mitigating the risks that autonomous systems pose to humans.

6.4.5 Using AI to Enhance Resilience and Safety

Unlike with a human operator, AI systems can analyse a huge amount of data and predict the consequences of a decision in a critical situation without suffering from the implication of human factors such as stress or environmental and organisational pressure. Even when not implemented as a fully autonomous system, Digital Twins and AI can be used to pass only the most relevant information with a clear level of "credibility" to a decision maker, cutting out some of the analysis fatigue for the operator. They could also be used to predict potential malfunctions and autonomously making proactive decisions.

In emergency response: ultra-fast decisions may be needed. Indeed in the future, through the use perhaps of quantum computing or surrogate models, one can posit that simulation and prediction might be possible on timescales faster that the physical phenomenon under investigation. This has the potential to allow considered actions to be taken to prevent or mitigate an accident quicker than a human decision maker could do, including in real-time.

Acknowledgement This work was supported by the Engineering and Physical Sciences Research Council, UK via a grant entitled: *Enhanced Methodologies for Advanced Nuclear System Safety (eMEANSS)*. The grant had reference: EP/T016329/1. The authors thank the EPSRC for this support.

References

Amram M, Kulatilkala N (1999) Project valuation using real options analysis. Havard Business School Press, Boston. ISBN 0-87584-845-1

Cardin MA (2013) Enabling flexibility in engineering systems: a taxonomy of procedures and a design framework. J Mechan Des 136(1):011005–011005. http://mechanicaldesign.asmedigit alcollection.asme.org/article.aspx?/https://doi.org/10.1115/1.4025704

Cardin MA, de Neufville RD, Dahlgren J (2007) Extracting value from uncertainty: a methodology for engineering systems design. In: 17th annual international symposium of the international council on systems engineering, INCOSE 2007—systems engineering: key to intelligent enterprises: Curran Associates, pp 1245–1259

Cardin MA, Zhang S, Nuttall WJ (2017) Strategic real option and flexibility analysis for nuclear power plants considering uncertainty in electricity demand and public acceptance. Energy Econ 64:226–237.https://doi.org/10.1016/j.eneco.2017.03.023

Copeland T, Antikarov V (2003) Real options: a practitioners guide: Thomson Texere. ISBN: 1-58799-186-1

D'Auria F, Mazzantinib O (2011) The Best-Estimate Plus Uncertainty (BEPU): challenges in the licensing of current generation of reactors. https://www-pub.iaea.org/MTCD/publications/PDF/ P1500_CD_Web/htm/pdf/topic4/4S08_F.%20D'Auria.pdf

de Neufville R, Scholtes S (2011) Flexibility in engineering design. Massachusetts Institute of Technology Press. ISBN: 978-0-262-01623-0

Estrada-Lugo HD, Santhosh TV, de Angelis M, Patelli E (2020) Resilience assessment of safety-critical systems with Credal networks. ESREL 2020 PSAM 15; Venice, Italy. https://pureportal.strath.ac.uk/en/publications/resilience-assessment-of-safety-critical-systems-with-credal-netw

ISO (2015) Quality management principles. https://www.iso.org/files/live/sites/isoorg/files/store/en/PUB100080.pdf

Locatelli G, Bingham C, Mancini M (2014) Small modular reactors: a comprehensive overview of their economics and strategic aspects. Progr Nucl Energy 73:75–85.https://doi.org/10.1016/j.pnucene.2014.01.010

Locatelli G, Fiordaliso A, Boarin S, Ricotti ME (2017) Cogeneration: an option to facilitate load following in small modular reactors. Progr Nucl Energy 97:153–161.https://doi.org/10.1016/j.pnucene.2016.12.012

Medkour M, Bouzaouit A, Khochmane L, Bennis O (2017) Transformation of fault trees into bayesian networks methodology for fault diagnosis. Mechanika 23 (6):891–899. https://doi.org/10.5755/j01.mech.23.6.17281

MIT Press (2024) Engineering systems: meeting human needs in a complex technological world. https://mitpress.mit.edu/9780262529945/engineering-systems/ Accessed Date 1 Aug 2024

Moore RL (2024) A reliable plant is a safe plant is a cost-effective plant: life cycle institute. https://www.lce.com/resources/a-reliable-plant-is-a-safe-plant-is-a-cost-effective-plant/ Accessed Date 21 Nov 2024

NREFS (2024) NREFS: management of nuclear risk issues: environmental, financial and safety. http://www.nrefs.org/. Accessed Date 10 Aug 2024

Petri CA (1962) Kommunication mit Automaten: Technischen Hoschule Darmstadt

Santhosh TV, Patelli E (2021) A resilience evaluation framework for complex and critical systems. https://cmswebonline.com/esrel2021-epro/pdf/357.pdf. Accessed Date 8 Nov 2024

Strategic Engineering (2024) Build a sustainable and resilient future. By design. https://strategic-engineering.co/. Accessed Date 16 Nov 2024

Shi HY, Song HT (2013) Applying the real option approach on nuclear power project decision making. Energy Procedia 39:193–198. https://doi.org/10.1016/j.egypro.2013.07.206

Systems Dynamics Society (2024) https://systemdynamics.org/ Accessed Date 21 Nov 2024

Trigeorgis L (1996) Real options, management options and strategy in resource allocation, 1st edn. MIT Press, Cambridge. ISBN 0-262-20102-x

United States Nuclear Regulatory Commission (2024) NPP PRA and RIDM: Early History. Lecture slides. Available at: https://www.nrc.gov/docs/ML1901/ML19011A423.pdf (accessed 24 November 2024).

World Nuclear News (2024) Guidelines drawn up for AI use in nuclear sector. https://www.world-nuclear-news.org/articles/guidelines-drawn-up-for-ai-use-in-nuclear-sector#:~:text=The%20UK%2C%20US%20and%20Canadian,maintaining%20adequate%20safety%20and%20security.&text=%22AI%20could%20benefit%20nuclear%20safety,of%20ways%2C%22%20they%20say. Accessed Date 22 Oct 2024

Chapter 7
Closing Remarks

7.1 Commentary

This book has examined various aspects of uncertainty handling and management relating to safety critical engineered systems. The book is not a history of engineering safety nor is it a textbook summarising and teaching established methods. It is closer to being a primer that describes and introduces a series of, generally, IT enabled approaches to data handling relating to safety critical or reliability concerns. The book does not dwell upon established notions in nuclear power plant safety such as the power and utility of ALARP[1] as a philosophy. Numerous excellent books and reports already exist on that topic. We take a different approach. The book is not grounded in safety, nor is it actually focussed on nuclear power. It is a book centred on uncertainty and how the study of uncertainty has shifted in recent years. Much of the innovation has arisen as a result of improvements in computing, but some advances have been more philosophical in nature—such as Bayesian sampling, or the limits of determinism.

While epistemic breakthroughs surely remain possible, one area of looming change is manifest. We are on the cusp of a revolution relating to the use of generative AI. This relatively new technology will be able to reveal in data, even in sparse data, that have eluded human analysts. We suspect that if a second edition of this book is ever produced, it will be a somewhat longer book than the version you have today.

---

[1] As Low as Reasonably Practicable.

© The Author(s) 2025
W. Nuttall et al., *Perspectives on Engineering Uncertainty*,
https://doi.org/10.1007/978-3-031-83254-3_7

7.2   Wider Applications

The purpose of this book is to share insights into, and pointers toward, uncertainty handling in nuclear reactor safety. While the methods have been developed in the context of nuclear reactor safety, and the example applications in this book are frequently focused on nuclear power stations, the analytical insights and techniques are not limited to that one field. We shall here consider its application to a number of other technologies.

7.2.1   Concerning Radioactive Waste

The dominant paradigm for high and intermediate level radioactive waste disposal is deep geological disposal. The two types of waste present very different engineering challenges and hence would not be housed together in the same underground galleries. In policy terms, and public acceptance terms, however, the two problems are not so far apart. Low Level Waste on the other hand has been a quiet policy success story in the UK with significant progress having been made over the last 20 years.

In the UK we have problematic, at a policy level at least, legacy wastes in the form of stainless steel barrels of intermediate level waste (ILW) in cementitious grout and vitrified High Level Waste (HLW) arising from past reprocessing. Formally, the key consideration relating to the definition of HLW in the UK "is waste where the temperature may rise significantly because of their radioactivity" (Nuclear Decomissioning Agency 2024).

One can imagine the design of a deep repository to have these two types of waste in separate galleries. Then national policy evolves and spent nuclear fuel is regarded as a waste for disposal. Spent fuel could be disposed of in the repository encased in copper canisters. The spent fuel is thermally hot, so at some level it is a kind of HLW, but its physical and chemical form is quite different from the reprocessing glass waste form. Formally the relationship between spent fuel and HLW in the UK is somewhat complicated. In the era of commercial nuclear fuel reprocessing (now ended) the UK made a clear distinction between spent fuel and waste, despite the fact that otherwise spent fuel meets the basic definition of HLW with self-heating arising from radioactivity. The logic against it being a waste was that spent fuel could be recycled into materials of commercial value (such as reprocessed uranium) although the strength of that logic could be challenged. Since the end of reprocessing some UK spent fuel has now been designated as waste and as such falls within the HLW inventory.

As spent fuel becomes regarded as a waste for disposal, then the focus turns the research that originally justified the repository concept which may have had in mind the older waste forms. In the UK no geological nuclear waste repository has yet been designed or built. That task lies in the future. The question of relevance to this book is: how does the shifting definition of waste relate to conceptions of uncertainty and flexibility in repository design (NEA/OECD 2012)? One issue of flexibility in the face of uncertainty is the notion of monitored retrievability around deep geological disposal. This has the benefit of allowing for easier intervention in the event that new fundamental truths emerge or become apparent, but equally the notion of 'keeping the repository doors open' erodes some of the safety case—for example if there is an air path from the surface environment to the geologically stored waste forms.

Much UK policy for radioactive waste focusses on the unavoidable question of legacy waste management, but here is also a need to consider future waste arisings, especially in the context of purported 'nuclear renaissance'.

Looking to the future of the nuclear fuel cycle, it will be necessary to understand what characteristics will be demanded from the next generation of fuels. How they can be manufactured, and their performance optimised? Similar considerations will also apply to structural materials in a nuclear power station. High-temperature reactors of the type being investigated in the UK will require many new materials. Also, knowledge of the physics of the nuclear reactions which provide the energy and the associated parameters, as discussed in Sect. 5.3, would need to be improved. The drivers motivating the further development of uncertainty thinking in nuclear engineering are there. The scope extends far beyond reactor design, construction and operation.

7.2.2 Concerning Space Exploration

After a 50 year delay, humanity is again taking staccato steps towards being a space-faring species. In the near to medium-term we expect to see real integration of complicated space systems. These will launch humans to the Moon and then possibly to Mars and beyond. Such systems must ensure life-support and system safety in a harsh environment. Wider considerations also include: security, energy provision, hazard avoidance and various human factors arising from physical and mental stress. These diverse issues must all be addressed within one functioning system.

Balancing energy demands for sustaining life with other mission needs has traditionally been carried out via iterative computations increasingly supported by computational models. Today we can build sophisticated uncertainty management into these models. As a result, old fashioned design conservatism can now largely be avoided.

It is important to be able to respond well to an unexpected event, such as the one that jeopardized the return to Earth of the Apollo 13 crew in 1970. Lead Flight Director, Gene Kranz, never actually said "failure is not an option", but the phrase became so associated with his leadership in the crisis that he chose it for the title of his 2009 memoir (Hagerty 2023).

Kranz had been trained to appreciate that the unexpected was to be expected. Now with our developed methods for uncertainty management, the arrival of modern computing and the advent of artificial intelligence with machine learning, we can conceive of a future, where complicated, life-sustaining systems can be operated safely, without direct human intervention. The Moon is a challenging environment where logistical options are limited. For lunar exploration and settlement to be a success it will be vital to manage and minimize key system uncertainties.

Readers might imagine that the issues of lunar missions and nuclear energy generation are entirely separate case studies, but it is becoming increasingly likely that there will be a role for nuclear energy on the Moon (Rolls-Royce 2023).

7.2.3 Concerning Railway Engineering

For this next example we come back down to Earth and consider an engineering sector that we have mentioned more than once in this book already. Chief Executive of the Office of Rail Regulation has claimed a reliably operating railway is a sign of a safe railway (Price 2013). Does it follow that a railway network that has loads of cancellations and late trains has a worse safety record than one that is more reliable in its operations? Work by Wemakor and colleagues (Wekamor et al. 2018) suggests not. Wemakor et al. were unable to find a strong correlation between operational and safety performance for UK train operating companies, but of course in such a single country study the fixed rail infrastructure (tracks etc.) is common to all operators. Consequently, if there is in the future an observed correlation between safety and reliability, then is it just a statistical accident that a railway system that manages to keep to the timetable and to have few cancellations is also statistically safer or should we look to limitations in the academic prior art?

To unpack such questions requires careful uncertainty analysis of the type explored in this book. In the world of sparse data, one must always be open to the question: *is it just a coincidence?* We are lucky indeed if we can get to the question—*are we dealing with correlation, not causation?* We note, for example, that correlation could be down to a common cause—such as good management.

- **The benefit of simplicity**

Railway engineering is replete with examples where the more modest incremental approach won out over a more ambitious and radical innovation. For example, in France the failure of the Aerotrain (Rail Target 2023) in the 1950s–1970s to its more conventional rival the Train Grandes Vitesse (TGV) is not widely remembered today, beyond a few railway experts. In the UK another competition between design philosophies is worth consideration. In the early 1980s there were two competing designs for high-speed rail on conventional UK track. The more ambitious was the Advanced Passenger Train (APT) developed by British Rail, and discussed earlier in Chap. 6 (page 72). The project was running late and despite eventually overcoming all technical obstacles, it was overshadowed by the roll-out across the UK rail network of the much more conventional high speed diesel locomotive the InterCity125. While 3 units of APT were deployed briefly in the mid-1980s, the InterCity125 would see more than 95 sets built in the period 1976–82 and history now judges that project to have been a success (Channel 5 2018).[2]

7.2.4 Concerning Automobile Manufacturing

Let us briefly consider the role of uncertainty in vehicle manufacture, particularly the Toyota production system and lean production: Prof W. Edwards Deming and others went to Japan from the US just after the Second World War and there they saw the early stages of the Japanese car industry. The emergent idea was that there would be a production line, and anybody could stop the production line. At the Ford Motor Company you could find that you got fired if you stopped the production line. The most important thing at Ford's was never to stop the line. But Deming went to Toyota, where the Japanese already had the basics of a good idea—stopping the line could improve production.

Imagine if anyone can stop the line. If as a factory worker, one sees that the painting operation is wrong, or if one sees that it's hard to put these bolts on, then one should stop the line. "Why is the paint so strange?"; "why can't I tighten the bolts?" and "why do the wipers never fit?" become stages in product improvement. One can stop the line and then the relevant team must have a proper conversation about why the process is not working well.

The production line only starts again when everybody is happy that the problem is fixed, and lessons have been learned. One might think that such an approach means the line will never run, because various people are stopping it all the time. Indeed, it might be stopped quite a lot at the beginning, but after a while the factory starts to produce a higher quality product consistently.

Once again we find ourselves thinking about the correlation, causation and sparse data. Did Toyota come close to defect-free manufacture because they had everything systematized or was there something else in play? Of course, low defect production, is not the same as reliability which, in turn, is not the same thing as safety, but we suggest that there is a connection between all three ideas and it is an interesting, if hard to formalise, connection.

At this point we approach the end of our book and we must simply retreat to the notion that the focus of this book is uncertainty and from there express our hope to the notion that the focus of this book is uncertainty and we hope that in these few small chapters we have been able to introduce some new perspectives and to present some opportunities for further study.

Acknowledgement This work was supported by the Engineering and Physical Sciences Research Council, UK via a grant entitled: *Enhanced Methodologies for Advanced Nuclear System Safety (eMEANSS)*. The grant had reference: EP/T016329/1. The authors thank the EPSRC for this support.

---

[2] It is interesting to note that commercial tilting trains were finally successfully deployed on the UK West Coast Mainline in 2002 in the form of British Rail Class 390 Pendolino rolling stock, as built by Alsthom and based upon Italian tilting train technology.

References

Channel 5 (2018) Intercity 125: the train that saved Britain's Railways, television documentary first broadcast 15 May 2018. Available at: https://www.youtube.com/watch?v=22VawS-FRjs

Hagerty M (2023) Gene Kranz never said 'failure is not an option' but his real legacy is 'tough and competent'. Houston Matters, Houston PBS. Available at: https://www.houstonpublicmedia.org/articles/shows/houston-matters/2023/08/29/460817/gene-kranz-never-said-failure-is-not-an-option-but-his-real-legacy-is-tough-and-competent. Accessed 9 Apr 2024

Nuclear Decomissioning Agency. UK Radioactive Waste Inventory. https://ukinventory.nda.gov.uk/about-radioactive-waste/what-is-radioactivity/what-are-the-main-waste-categories/. Accessed Date 17 Nov 2024

NEA/OECD (ed) (2012) Reversibility and retrievability in planning for geological disposal of radioactive waste: proceedings of the "R&R" international conference and dialogue

Price R (2013) An efficient railway is a safe railway. IOSH Keynote Speech. https://www.orr.gov.uk/media/10832/download

Rail Target (2023) Jean Bertins Aeortrain: the story of a French genius and his failure. https://www.railtarget.eu/technologies-and-infrastructure/jean-bertins-aerotrain-the-story-of-a-french-genius-and-his-failure-5840.html. Accessed 9 Apr 2025

Rolls-Royce (2023) Media release. Rolls-Royce unveils Space micro-reactor model for moon exploration. Available at: https://www.rolls-royce.com/media/our-stories/discover/2023/rr-unveils-space-micro-reactor-model-for-moon-exploration.aspx

Wekamor W, Anson J, Schmid F (2018) Establishing the relationship between railway safety and operational performance. Int J Transp Dev Integr 1:98–114. https://www.researchgate.net/publication/325747705_Establishing_the_relationship_between_railway_safety_and_operational_performance

Wikipedia. InterCity 125. https://en.wikipedia.org/wiki/InterCity_125. Accessed Date 22 Oct 2024

Glossary of Key Terms

Absorption A fraction of the neutrons, rather than going on to induce further fission, are absorbed by other elements in the reactor. These absorbers include the fuel, cladding or the moderator, other absorption can occur within the reactor components designed to control the reactor, such as control rods.

ALARP As low as reasonably practical. A principle used in risk management. Judgement as to what constitutes "reasonably practical" necessitates weighing a risk against the trouble, time and money needed to control it.

Aleatory Uncertainty Aleatory Uncertainty addresses inherent variability in systems that cannot be eliminated even with complete knowledge. Also known as irreducible uncertainty, aleatory uncertainty is attributed to inherent randomness or variability in natural phenomena. It is typically modelled using probability distributions to describe the range of potential outcomes.

Alpha Cut An alpha cut is a horizontal 'slice' taken from the cumulative distribution function (CDF) of a Probability Box (P-box), at a value between 0 and 1. This creates an interval between the two intersections of the line and the two CDFs of the P-box.

Artificial Intelligence/Machine Learning Artificial intelligence refers to the general ability of computers to emulate human thought and perform tasks in real-world environments. Machine learning refers to the technologies and algorithms that enable systems to identify patterns, make decisions, and improve themselves through experience and data.

Attractor The state, or value, to which a dynamic system tends to move.

Canonical As in forming part of a 'canon'. A canon is a widely accepted set of works that are central and core to a discipline of study. As such, canonical can mean archetypical or according to accepted wisdom in the field of investigation.

Cascading Failure Failures in a system, comprised of separate components, where failure in one or more components leads to failure in other components and then failures in yet more components.

Correlation The relationship between changes in two variables. Numerically a correlation of 1 means that a change in one variable is fully reflected in the other. Conversely a correlation of 0 means that the changes in the two variables are totally independent.

Criticality/Reactivity Depending on the composition and configuration within the nuclear reactor the proportion of released neutrons that go on to cause further fissions will vary. This in turn results in variation in the number of fissions occurring in each generation of induced fission. If the ratio of the number of fissions is equal to the number in the previous generation then the reactor is said to be critical. If it is less, the reactor is sub-critical, and if it is greater, the reactor is super-critical.

This state can also be quantified by the reactivity. A reactivity of 1 represents a critical state, a value less than 1 sub-criticality and a value greater than 1 super-criticality.

Cumulative Distribution Function (CDF) See Probability Distribution below.

Decay Heat Heat that is produced in the fuel after the reactor has been shut down. The heat arises as a consequence of the decay of radioactive fission products within the fuel.

Decision Tree A form of flowchart used to visualise a series of possible courses of events with a branching structure.

Deduction A reasoning process which seeks to arrive at a valid conclusion from a set of initial premises (c.f. induction).

Deterministic An approach to calculations in which no account is taken of uncertainties. This results in a single value for the output with no range of possible values.

Digital twin A virtual model of a physical object, or system, intended to serve as a counterpart to the physical object, responding to inputs in the same way as the physical object.

Double Loop Sampling This is a two-stage approach to sampling. An initial sampling process is carried out, and the results are analysed. The results of this analysis are used to specify a second sampling process.

Engineering Model A set of mathematical equations (either explicitly, or within a piece of commercial software and hence unseen by the user) which characterises an engineered system by connecting a series of input parameters (i.e., physical loading, temperature, geometry, etc.) to a series of outputs of interest (i.e., physical strain/stress, electrical output, emissions, etc.).

Epistemic Uncertainty Epistemic uncertainty deals with uncertainties arising from incomplete knowledge or lack of information about a system. This type of uncertainty is often considered reducible through additional data collection, research, or improved modelling techniques.

Fissile Material and Enrichment Only a small number of isotopes are able to undergo fission at thermal neutron energies, the ones of central importance being U-235 and Pu-239. In many reactor designs, including the AGR the uranium must be enriched, i.e. the proportion of U-235 increased above the naturally occurring proportion of 0.7%. The necessary amount of moderation varies between reactor designs.

Fission When a nucleus of a fissile material is struck by a neutron, it may split into two or more nuclei, releasing energy. The splitting may also release neutrons, which in turn may strike other nuclei, causing them to split. This can lead to a so-called "chain reaction" of fissions. The fission of the nuclei also results in fragments from the original known as fission products. Not all of the neutrons will go on to cause further fissions. Some will be absorbed by reactor components, and the surroundings. The energy released during fission heats up the fuel, is removed by the coolant system and is (as steam) used to power turbogenerators which produce electricity.

Flux/Fluence Flux is a measure of the numbers of particles (e.g. neutrons or gamma particles) that pass-through a given area in a specified period of time. The total number of particles that have passed through the area is the fluence.

Fuel/Core The fissile material, together with other materials (including non-fissionable isotopes of the fissile element) is fabricated into to fuel elements. These elements, together with absorber control rods and graphite moderator blocks are combined into the reactor core.

Fuel Stringer The AGR fuel element comprises enriched uranium oxide pellets stacked inside stainless steel tubes. The tubes are assembled in a graphite sleeve forming a fuel assembly. Each AGR fuel assembly is made of 36 steel tubes, each containing 64 pellets. An AGR Fuel Stringer is a set of 8 fuel elements held end-to-end hanging together on a tie bar,

Gamma Radiation Gamma radiation comprises high energy photons (light "particles") which can be released during fission or the decay of radioactive material. They differ from X-rays in that gamma particles come from within the nucleus.

Induction A reasoning process which seeks to derive general conclusions from individual instances or observations (c.f. induction).

Isotope Isotopes are different forms of the same chemical element with different numbers of neutrons in the atomic nucleus. The nuclear properties of different isotopes may be very different. Different isotopes are distinguished by adding a superscript to the chemical symbol for atomic mass of the element, e.g. ^{238}U and ^{235}U refer to the isotopes of Uranium containing a total of 238 and 235 protons and neutrons in the nucleus respectively. In this book we may use an alternate notation: U-235 and U-238.

Mathematical Model A quantitative description of an object or process making use of mathematical equations.

Moderation A neutron is more likely to induce fission when travelling at speeds lower than those at which they have immediately after they have been released in the fission process. This is achieved through a series of collisions between the, initially fast, neutrons and the atoms in material such as water or graphite. This process of slowing to 'thermal energies' is known as moderation. A neutron which has been slowed down by the moderation process is often referred to as a thermal neutron.

Monte Carlo Modelling A technique that uses repeated random sampling of the input parameter values to produce a range of values for the model output.

Neutron Neutrons are fundamental particles which, together with protons, comprise the atomic nucleus. Neutrons have no electrical charge and can be released during fission.

Primary/Secondary Circuit The primary circuit contains the coolant that passes through the reactor core. Heat is transferred (in the steam generators) to the secondary circuit in which steam is produced to drive turbogenerators for the generation of electricity in a power station.

Probability Distribution/Cumulative Distribution Function The probability distribution function describes how the value of a variable is distributed across its range of possible values. A Cumulative Distribution Function gives the same information, but in terms of the probability that the value is less to or equal to the specified level. Note Probability Distribution Function should not be confused with Probability Density Function (PDF), that is the special case applicable to a continuous random variable.

Quantum Computing A computing technique that makes use of quantum mechanical phenomena. The use of these phenomena means that some calculations can be carried out significantly quicker than can be done using "classical" computers.

Reliability The probability that a product, system or service will perform its function adequately for a defined period of time or will operate in an environment without failure.

Resilience The ability of a system to respond to adverse impacts and remain in, or quickly return to, a stable state.

Safety Margin A safety margin (or margin of safety) is taken in this book to mean an intentional overestimation of certain parameters to ensure that a design or system can tolerate unexpected variations or uncertainties without compromising safety or functionality.

Sensitivity A measure of how much the output of a computer model changes as a result of a change in the value of an input parameter.

Surrogate Model A simplified approximation of a more complex, higher-order model. These are often used when use of the high-level model would be resource intensive.

TRISO Fuel TRI-Structural Isotropic fuel, a form of nuclear fuel comprised of an enriched (usually circa 19%) uranium oxide kernel surrounded by layers of carbon and silicon carbide to prevent the release of fission products.

Validation The process of comparing the output of a mathematical model against data obtained from observation of the physical system that the model represents.

Index

© The Editor(s) (if applicable) and The Author(s) 2025
W. Nuttall et al., *Perspectives on Engineering Uncertainty*,
https://doi.org/10.1007/978-3-031-83254-3